Curve Ball

Baseball, Statistics, and the
Role of Chance in the Game

Jim Albert

Jay Bennett

COPERNICUS BOOKS

An Imprint of Springer-Verlag

Published in the United States by Copernicus Books,
an imprint of Springer-Verlag New York, Inc.
A member of BertelsmannSpringer Science+Business Media GmbH

Copernicus Books
37 East 7th Street
New York, NY 10003
www.copernicusbooks.com

Library of Congress Cataloging-in-Publication Data
Albert, Jim
   Curve ball : baseball, statistics, and the role of chance in the
   game / Jim Albert, Jay Bennett.
       p.   cm.
   ISBN 0-387-98816-5 (alk. paper)
      1. Baseball—Statistics.   I. Bennett, Jay.   II. Title.
GV877.A46 2001
796.357′01′5195—dc21   2001017397

Manufactured in the United States of America.
Printed on acid-free paper.

9 8 7 6 5 4 3 2 1

ISBN 0-387-98816-5   SPIN 10717015

Contents in Brief

Contents

Introduction

The public seems to have an insatiable appetite for baseball statistics. Evidence for this assertion can be found everywhere. Baseball statistics in the form of box scores, lists of player batting averages, and compilations of pitcher records are published (and revised daily) in newspapers and on websites. Sportscasters on television and radio quote them constantly. Callers to talk radio shows (many of whom probably grudgingly endured math classes in school) cite them selectively to make a point. Baseball encyclopedias, annuals, and yearbooks provide row after row of numbers through which baseball fans can wade (much as Kevin Costner waded through the corn in *Field of Dreams*), imbibing the game through statistics. A true fan can disappear into these books for hours on end to emerge freshened and enlightened. Even a film as emotional and nostalgic as *Field of Dreams* featured Costner quoting the statistics of Shoeless Joe Jackson, and had James Earl Jones presenting an oration on the meaning of baseball while cradling a baseball encyclopedia in his arms.

The uses of statistics in baseball vary. Major League Baseball, of course, is first and foremost a form of entertainment. Statistics on the players and teams are used to supplement the fans' enjoyment. Sportscasters broadcasting a game cite the pitcher's and batter's statistics to heighten the suspense of the confrontation between the two and the possible strategies each might use to gain an edge. And so it is that almost any situation that arises in the game presents an opportunity for the color analyst to cite a related statistic:

1. With an excellent runner on first base, the fan is treated to statistics on the runner's percentage of success in stealing second base, the pitcher's ability to prevent the runner from attempting the steal, or the catcher's success in gunning down runners attempting to steal. Often the viewer or listener will be treated to all three.

2. When a closing reliever enters the game in the late innings to preserve a victory, it is *de rigueur* for the sportscaster to inform the listener or viewer of the number of times the pitcher has succeeded and failed in saving the game. Also often required: data on how good the team is at holding on to the lead.

3. The appearance of a pinch hitter immediately elicits a brief history of the new batter's success against the pitcher, and against comparable pitchers, and against pitchers in this kind of situation. . . .

The leisurely pace of a baseball game not only makes such recitations possible, it makes them desirable to many fans of the game, an aid to increasing drama and sustaining interest.

And it's not only fans who have an interest in statistics. Baseball is both a vocation for professional athletes and a leisure activity for amateurs. But in either case, it is a highly competitive and highly measurable activity, and it is only human nature for those who are involved to want to know exactly how well they are doing. Baseball statistics fulfill this "yardstick" role for players, of course, but also for a team's management (owners, managers, and coaches, for example). They use statistics in making strategy decisions on the field, in planning player development programs, and in putting together teams. And on the business side of the professional game, certainly, baseball statistics provide ammunition in salary negotiations.

Off the field, gamblers have long been interested in baseball, today making bets to the tune of $30 million dollars annually (*Total Baseball,* p. 667). Analyzing the statistics of the game to gain an edge is a necessity in wagering. While football overtook baseball in the gambling arena (with three times the amount wagered on baseball in recent years), baseball has provided a gaming venue that is totally dependent on statistics—no actual play required! In the early 1980s, a small group of baseball enthusiasts began to gather annually at a restaurant in Manhattan (*La Rotisserie Française*) to draft teams for every new baseball season. While the names of the players and their statistics were all real, the teams and the league itself were a complete illusion, the dream of fans who wanted to act out the fantasy of being a major-league general manager. The basic game was (and is) very simple. Each player drafts a team of baseball players, taking names from the 30 major-league teams. Throughout the year, the actual statistics for the players are counted for the fantasy teams that drafted them. At the end of the season, using a variety of scoring algorithms, the team with the best stats is declared champion. *Rotisserie Baseball* (or *Fantasy League Baseball*) has so expanded in popularity that next to stock market analysis, it may be the most widespread application of statistics in the United States. The concept has expanded to Fantasy League Football, Fantasy League Basketball, even Fantasy League NASCAR Racing (so we are told). To us, at least, this is a unique idea: statistics spawning an entertainment industry!

Unlike other major sports, in which interest in statistics is confined almost exclusively to the season in which the games are being played, baseball discussions appear in the media throughout the year. In a sense, the baseball off-season, with its own schedule of awards and winter meetings leading up to spring training, has a life of its own. The merits of various awards (Cy Young, MVP, and Rookie of the Year) and honors (election to the Hall of Fame) are debated annually in statistical terms. (Bill James wrote an entire book, *The Politics of Glory*, on the subject of Hall of Fame election.) Off-season trades are evaluated with reference to player statistics. Multiyear, multimilllion-dollar contracts are analyzed in terms of stats. Clearly, it is statistics that provide the fuel for hot-stove leagues (which have expanded from the neighborhood tavern to talk radio). In the dead of winter, it might be said, statistics can keep the game alive.

Given our inundation in baseball statistics, given the year-round attention thousands devote to the topic, one may wonder what in the world we hope to add in this publication. To put it simply, we wish to contribute a statistician's perspective to this massive collection of data. This statement may give you pause. You may ask who is collecting all this data except statisticians, and once the data have been collected, sorted, and totaled—clearly, something that's already been accomplished—isn't the statistician's job done? Actually, for a statistician, once the data have been compiled (whether it's baseball data, medical data, education data, whatever), the job has only started. It is the statistician's aim to understand the underlying structure in a data set and to elicit the truths hidden within it. The inspiration and imagination needed to accomplish this requires a kind of art from the statistician, a reason why the *pro forma* production of pie charts, bar charts, and *t*-tests is often ineffective in statistics, and indicative of poor statistical analysis.

At this point, we have to make a difficult and maybe boastful-sounding distinction in order to express what we mean when we talk about a statistician's perspective. It may seem paradoxical, but most sports statisticians are not statisticians. We know that sounds like most chocolate candy is not candy, but bear with us. We use the term "sports statisticians" to refer to those people who work for sports teams, sports leagues, and media covering sporting events and whose main job is to record data from the sporting event, summarize the data in totals and averages, monitor records, and identify interesting patterns in the data that might be used by broadcasters or by teams in bargaining with players. Although somewhat dated now, in 1978, Arthur Friedman, a sports statistician for many New York teams including the Mets, Rangers, Knicks, and Jets, documented the basics of the job in *The World of Sports Statistics*. We guarantee that you won't find any references to linear regression or standard deviation in Friedman's description of the world he inhabits. Yet these are among the most basic tools and terms taught in statistics courses in high school and college, and the on-the-job tools of the professional statistician.

Sports statisticians do an excellent job of addressing the information needs of their audience: management, athletes, broadcasters, and fans. However, this audience is unaware of or indifferent to the fact that statistical analysis can be performed at a higher level—and one that can be very rewarding because it can lead to new levels of understanding. We are reminded of a story once told by Bud Goode, a successful consultant for the NFL and a syndicated columnist on sports statistics. While he had great success in selling his services to the NFL, Goode found great resistance in persuading baseball management of the value of his work. Meeting with the Los Angeles Dodgers, he told Walter Alston, their Hall of Fame manager, that through his statistical analysis he could gain the Dodgers half a run per game—a sizable increment. Alston described this wondrous promised result to one of his coaches, Danny Ozark, who responded, "How do you score half a run?" Now, Ozark was never known as a deep thinker about numbers. "Half this game is 90 percent mental," is only one of a full page of Ozarkisms from *Baseball's Greatest Quotations*. Still, it does give a picture of the circumscribed horizon of many people toward more sophisticated uses of statistical techniques. To paraphrase another Ozarkism, too often this audience's limits are limitless when it comes to statistics.

Because of the limited demands of fans, players, and management, sports statisticians have not found it necessary to employ or even be trained in standard statistical applications or theories. As Goode discovered, being too sophisticated about the subject can actually be a deterrent. Nonetheless, innovative statistical research has been and continues to be done with baseball data. At first, professional statisticians, interested in sports statistics but left unsatisfied by the analyses in the national media, began exploring baseball data on their own. Then, in the 1950s, out of their love for baseball and statistics and for the sheer fun of it, professionals began to apply their skills to baseball data. Soon their results were published in conference journals and in professional journals.

Only rarely have these works caught the attention of the media and thereby the public. In the past, the majority of professional statistics colleagues looked somewhat askance at these efforts, not unlike parents who chastised their children for ignoring their studies to play baseball (in past generations, of course, before ballplayers were multimillionaires). Recently, however, this attitude has changed. Along with other scientific pursuits, the statistics profession has found itself somewhat isolated from the public, to whom its methods appear arcane and its goals at best a mystery. What better way to reconnect to the public (and especially the nation's youth) than through the application of statistics to sports? In the 1990s, the American Statistical Association created a section on Statistics in Sports, and the International Statistical Institute (ISI) created a Sports Statistics Committee.

As described in Jay Bennett's snapshot review of baseball statistical research in *Statistics in Sport,* advanced use of statistics has not been applied exclusively by professionals. Members of SABR, the Society of American Baseball Research (the root of the term "sabermetrics"), also recognized deficiencies in the presentation of baseball statistics. By means of its *Baseball Research Journal* and *By the Numbers* newsletter, a dedicated coterie of members have made intriguing analyses of baseball data. Several of these sabermetricians (Bill James, Pete Palmer, and Craig Wright, among others) have gone on to build careers out of the analysis of baseball statistics. The rigor and understanding of statistical theory displayed by sabermetricians (professional or non-professional) in their publications can vary wildly, from the level of a talk-radio telephone caller to that of a professional statistician. (Indeed, many professional statisticians, ourselves included, are members of SABR.) But the impulse to bring more sophistication to analysis of the data has certainly done a lot to broaden the appeal of statistics.

What differentiates the work of professional statisticians (and many sabermetricians) from that of sports statisticians? In short, sports statisticians do not apply statistical models to data. Their analysis generally consists of summing numbers, finding averages, making comparisons of these computed numbers, and perhaps inventing a new formula based on the raw data (Slugging Percentage being an early example). At the highest level, the work of the sports statistician would be described by a professional statistician as exploratory data analysis, looking for patterns in baseball statistics. For professionals, statistical models are the key. They provide the means of getting to the truth behind the data. By applying statistical models, professional statisticians can perform confirmatory data analysis, calculating the degree of confidence to which a pattern or statement can be said to be true.

The primary goal of this book is to provide insights that can be gleaned when statistical models are applied to Major League Baseball data. In our own research and in the work of others, we have found that this type of investigation sheds new light on the game, and increases our admiration for those who participate in it, and especially those who excel. Our research has also given us an increased appreciation for the power of advanced statistical techniques, and thus the secondary objective of the book: to convey this appreciation for statistics to the reader. The book does not just recite numbers, provide lists of players, and recount a litany of astounding results; it also describes the logic that leads us to conclusions that have a statistical basis. In this way, we hope that the reader will gain a better understanding of statistics in general.

Many of you are probably more familiar with statistical models than you realize. One of the more common examples is a simulation model. Tabletop baseball board games use simple statistical models as their foundations. So, it is only fitting Chapter 1 is in large part a discussion of these games, which have been familiar to many of us since childhood. Chapter 2 moves on to a brief overview of some common baseball statistics, but presented from a new perspective. The chapter will introduce a number of data analysis concepts that will prove to be useful in the remainder of the book.

In the next set of chapters, we make the crucial distinction between ability and performance. Chapter 3 introduces the distinctions between these concepts with respect to getting on base. These basic concepts are then extended to examine two issues much discussed among baseball fans. Chapter 4 looks at the significance of breakdown statistics for different batting situations (e.g., facing a righty pitcher versus a lefty, hitting at home versus away). These numerical darlings of broadcasters are the bread-and-butter statistics of pre-game and post-game shows. However, the restricted nature of the categories often leads to small sample sizes, leading us to ask whether the observed differences are truly significant or just the product of chance. Another favorite of broadcasters is the batting streak, which we focus on in the next chapter. Statistics are used to identify who is hot and who is not in a given game (or week or month or season), and some players are identified as being generally streaky hitters. They do not hit equally well over a season, instead tending toward hot spells and cold spells. Chapter 5 examines how we might determine whether a player is a streaky hitter, or whether these streaks are random occurrences.

In the middle chapters, we shift gears a bit. The basic subject remains batting, but where earlier chapters stuck to traditional measures of hitting like Batting Average, we now examine some of the alternatives that have been suggested by researchers in recent years. Chapter 6 presents a thorough description of how we can compare these measures and provides a test for the significance of their differences. Most of these alternatives were developed from an intuitive understanding of run production, in which the researcher starts with a "common sense" explanation for how teams get runs. In Chapter 7, we look at some models that use the data from baseball history to develop measures of batting performance. Chapter 8 starts with yet another approach, presenting a model based not on intuition or data, but on a logical construction of the probabilities of scoring runs. This "simulation model" can be used to provide support for two simpler models, which are then compared in their capacity to measure the batting performance of individual players.

As we move into the late chapters, we start to consider those elements of the game that contribute to victory. In Chapter 9, we look at the concept of clutch hitting—that is, getting a hit at a critical moment, when the stakes for winning or losing a game are at their highest. While the existence of the ability to hit in the clutch has been much debated, a fan knows a clutch hit when he or she sees one. The question is, can the value of a clutch hit be quantified not just with respect to run production, but also as to how it actually contributes to winning the game?

Most of the chapters up to this point address different ways to evaluate what players or teams have done in the past, and how this past performance relates to ability. In Chapter 10, we examine how models can be used to make *predictions* on the seasonal performances of players and teams, as well the future career achievements of players. Chapter 11 makes a final statement about the influence of chance on the ultimate goal of every team, winning a World Championship. We always think of the World Series victor as the best team in baseball, but you may have a new sense of the role of chance by the end of this chapter.

It has been said that the primary difference between a successful minor-league hitter and a successful major-league hitter is the ability to hit the curve ball. All professional hitters (major or minor league) can hit a fastball when they are prepared for it. But in order to advance to the top level, the player must master the skill of hitting the curve. The same can be said of baseball statistics. The average fan gets a great deal of information on statistics from newspapers, magazines, television, radio, and web sites. But to see the truth behind the numbers, the fan's ability to analyze data must be raised to a higher level. He or she must master not just the records and averages printed on the sports page, but some of the models we describe in *Curve Ball,* and above all else, as we reiterate in Chapter 12, the role of chance in the game.

Acknowledgments

Each of us is lucky enough to have a partner who has been supportive and patient during the long process of researching and writing and producing this book. To our respectives wives, Anne Albert and Lynn Bennett, we owe thanks more than we can say.

We would also like to acknowledge David Bernstein, Jim Box, Michael Doherty, David Grabiner, David Jones, Kenneth Ross, and Bob Wardrop, as well as two anonymous reviewers, for their insightful comments and suggestions on our manuscript in the course of its development.

And we thank Paul Farrell and John Kimmel, our editors at Copernicus Books and Springer-Verlag, respectively, for their faith in our project and their support throughout its creation. Thanks also to Production Editor Mareike Paessler and Designer Jordan Rosenblum of Copernicus Books for their very thoughtful and helpful contributions.

Jim Albert
Jay Bennett

Publisher's Note

Additional information related to this book—including supplementary materials on tabletop baseball games, as well as a list of errata as we find them—can be found at www.copernicusbooks.com.

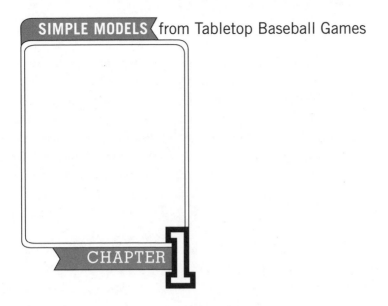

Two boys stared out the window at the rain-swept streets. It looked like a long afternoon with nothing to do. If this were today, the boys would have hurried to the nearest available TV or computer to play the latest electronic game, but this was the 1950s, when Nintendo could only have been another monster intent on devouring Tokyo. Baseball was still the dominant sport in America, and if you couldn't play baseball outdoors, the next best thing was baseball indoors.

All-Star Baseball (ASB)

The closest equivalent to *High Heat Baseball* back then was the Cadaco board game *All-Star Baseball (ASB)*. The boys opened the box and started dividing up the players into the American League and National League teams. The instantly recognizable players were placed in the starting lineups. Mickey Mantle, Willie Mays, Duke Snider, and Hank Aaron, all current stars, were snapped up immediately. Then there were the great old-time players their fathers had eulogized—legends like Babe Ruth and Lou Gehrig. These guys had to be included as well.

The lineups shaped up beautifully at first, but then choices started to become difficult. Other old-time players were not so well known. And many of the current players were familiar from baseball cards, but they were not titans, not obvious choices like Willie, Mickey, Hank, and the Duke. The game did not come with ready-made lineups or even tables of batting averages as current electronic games do. How were the boys to choose the remaining players?

Although they weren't conscious of it, what they did next was an intuitive form of data analysis. At the heart of All-Star Baseball were the disks representing the batting abilities of individual players. A rough replica of an ASB disk, in this case for the incomparable Babe Ruth, is shown in Figure 1-1. When the ASB version of the Babe came to bat, his disk was placed under a spinner. With a flick of the finger, the mighty batter swung, with the result determined by where the pointer came to rest (after reference to a chart on the giant scoreboard dominating the field). Most results are clearly understood from the legend in Figure 1-1; note, however, that GB and FB are ground-ball and fly-ball outs, respectively.

The pie slices on the disk were in effect a visual representation of Ruth's ability. Because they were experienced ASB players, the boys could quickly see why the Babe was everything their fathers had claimed. The slice labeled 1 may not look large, but compared to any other player's 1 slice, it was enormous. Are triples (slice 5) really that rare, or was Ruth just slow? (Of course they are rare.) And look at those two expansive 9 slices for walks. Pitchers were fearful of Ruth . . . with good cause.

Ruth was a player apart from all the others, and a natural choice for the starting lineup. Choosing other players was not so easy. What about a starting first baseman for the National League? Two candidates, Gil Hodges and Bill Terry, looked promising. Hodges was well known to the boys, an All-Star for the

BABE RUTH

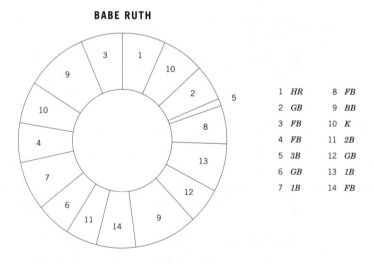

1	HR	8	FB
2	GB	9	BB
3	FB	10	K
4	FB	11	2B
5	3B	12	GB
6	GB	13	1B
7	1B	14	FB

FIGURE 1-1   Replica ASB disk for Babe Ruth.

powerful Brooklyn Dodgers. Terry was not familiar, having retired at the conclusion of the 1936 season, after 14 years with the New York Giants. To settle the question, the boys instinctively made a direct visual comparison of pie slices. Of course the first comparison was for the slice numbered 1, the one and only slice for home runs. Figure 1-2 presents a doughnut plot of the Hodges and Terry disks aligned for this comparison. (A doughnut plot places one pie chart inside another for ease of comparison.) Hodges has the definite HR edge (slice 1), but a closer look shows that, apart from power, Terry does have some real strengths. His slices for singles, doubles, and triples clearly dominate those of Hodges. Terry's walk slices are not as large as those of Hodges, but his strikeout areas are very small. (He averaged only about 30 strikeouts per season.) More or less intuitively, the boys saw that their choice came down to whether they needed a batter to get on base (Terry) or one to drive in runs (Hodges).

In this brief description of All-Star Baseball, we encountered two basic statistical concepts:

- *Visual presentation is a powerful tool for identifying patterns and making hypotheses.* To the boys, the ASB disks were extremely useful visual presentations, ideal for quickly assessing players and making lineup choices. In particular, ASB disks introduced the boys to *pie charts,* a graphic representation commonly used in statistics.

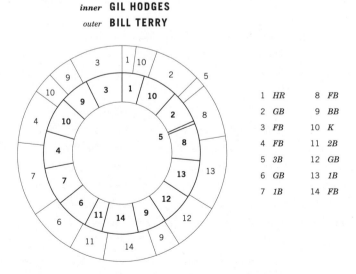

1	HR	8	FB
2	GB	9	BB
3	FB	10	K
4	FB	11	2B
5	3B	12	GB
6	GB	13	1B
7	1B	14	FB

FIGURE 1-2   Comparison of replica ASB disks for Gil Hodges and Bill Terry.

- *Data often must be converted into a more useful form before meaningful conclusions can be derived from it.* Apart from batting averages, baseball cards of the 1950s presented only event totals (for example, the number of home runs in a season or a career). Total numbers of home runs are important, but how many opportunities (plate appearances) were necessary to achieve those totals? For the purposes of the game, this baseball card data had been converted into a form more useful for modeling what actually happened on the field. The use of ASB disks (in the form of pie charts) emphasized the importance of examining player performance in terms of *proportions of opportunity* rather than mere count totals.[1]

Perhaps the most powerful concept presented by All-Star Baseball was its presentation of a relatively simple but effective statistical model of outcomes. Again, many of the aspects of this model were picked up by the boys intuitively. They understood that the spinner was a *randomizing device.* The outcome of a plate appearance was a random event affected only by the proportion of disk space occupied by each possible result. And this was the rationale for their comparison of slices on the disks: the larger the slice, the greater the chance of obtaining that result. (Of course, this did not stop the boys from doing all they could to consciously influence the outcome: incomplete spins and soft flicks from carefully chosen starting places prompted numerous protests and calls for "official" rulings.)

The disks presented in Figure 1-2 were created from the career statistics for Terry and Hodges as shown in Table 1-1. The first column for each player indicates the number of at-bats (AB), hits (H), doubles (2B), triples (3B), home runs (HR), walks (BB), and strikeouts (K) given for each player in the book *Total Baseball.* The number of singles (1B) can be determined by subtracting the numbers of doubles, triples, and home runs from the number of hits. The sum of at-bats and walks is used as the number of plate appearances (PA), ignoring relatively rare events such as sacrifice flies. The second column for each player presents the percentage of plate appearances in which the event occurs. The third column for each player translates each percentage into degrees of arc. The entire disk has 360 degrees. Thus, the arc spanned by each type of play is 360

---

[1] ASB disks were also more complete in their information. Walks were often not presented at all in baseball cards in the 1950s, while ASB disks provided a graphical representation of the ability to obtain walks (and thereby a more complete picture of the ability to get on base).

times the appropriate percentage. For example, a player who obtained 25 home runs in 500 plate appearances was able to hit a home run in 25/500 or 5 percent of his plate appearances. To represent this performance on an ASB disk, the HR slice (slice 1) would span 5 percent of 360 degrees, or 18 degrees.

The only characteristic of the disks which is not determined is the split of outs. Strikeouts are determined directly from the data as shown in Table 1-1, but there are no data on ground-ball and fly-ball outs. Past disks appear to assume that 60 percent of outs from batted balls are fly-ball outs (FB) while the remaining 40 percent are ground-ball outs (GB). This assumption was used in the construction of the replica disks in Figure 1-2 and could be used for any player from data in a standard baseball encyclopedia or on the reverse of a baseball card.

To use a technical statistical term, All-Star Baseball models the result of each plate appearance as an outcome from a *multinomial distribution*. A multinomial distribution has a finite set of possible outcomes, each with a fixed probability of occurrence. Each occurrence is a random result from this set, dependent only on the probabilities. In ASB, the set of all possible occurrences are enumerated on the playing field. Based on the individual disks, each batter has his own multinomial distribution defined by his unique set of probabilities. Table 1-2 presents the multinomial distribution represented by Bill Terry's disk. For each outcome,

	BILL TERRY			GIL HODGES		
Play	*Number*	*Percent*	*Degrees*	*Number*	*Percent*	*Degrees*
AB	6428	•	•	7030	•	•
H	2193	•	•	1921	•	•
1B	1554	22.3%	80	1208	15.2%	55
2B	373	5.4%	19	295	3.7%	13
3B	112	1.6%	6	48	0.6%	2
HR	154	2.2%	8	370	4.6%	17
BB	537	7.7%	28	943	11.8%	43
K	449	6.4%	23	1137	14.3%	51
FB	2272	32.6%	117	2383	29.9%	108
GB	1514	21.7%	78	1589	19.9%	72
PA	6965	100.0%	360	7973	100.0%	360

TABLE 1-1   Converting Baseball Data into All-Star Baseball Disks

DISK			MULTINOMIAL		BINOMIAL	
Result	Number	Degrees	Individual	Total	Probability	Result
1B	7	40.2	.1116			
1B	13	40.2	.1116	.223		
2B	11	19.3	.0536	.054		
3B	5	5.8	.0161	.016		
HR	1	8.0	.0221	.022		
BB	9	13.9	.0385			
BB	9	13.9	.0385	.077	.392	On-Base
K	10	11.6	.0322			
K	10	11.6	.0322	.064		
GB	2	26.1	.0725			
GB	6	26.1	.0725			
GB	12	26.1	.0725	.217		
FB	3	29.4	.0815			
FB	4	29.4	.0815			
FB	8	29.4	.0815			
FB	14	29.4	.0815	.326	.608	Out
		360	1	1	1	PA

TABLE 1-2   Conversion of Bill Terry's Replica ASB Disk into Multinomial and Binomial Probabilities

the table presents the arc (in degrees) spanned by the outcome on the disk. Some occurrences have more than one slice on the disk (undoubtedly to discourage creative spinning), so the probability is the sum of the arcs for each slice. The arcs are then converted to probabilities by dividing the arc degree values by 360 (the number of degrees in a full circle). The basis for this conversion is that when the spinner is struck, every direction or point on the disk's circumference has an equal chance of being the result—the point at which the spinner comes to rest. We can collapse the list of ASB results in Table 1-2 into two important categories, each consisting of several play results:

1. *On-Base*: the single, double, triple, home-run, and base-on-balls results.

2. *Out*: the fly-out, ground-out, and strikeout results.

BILL TERRY

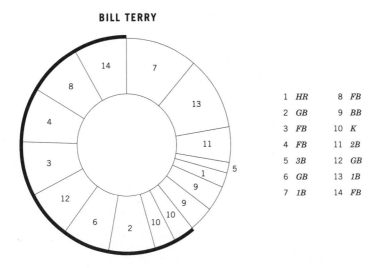

1	HR	8	FB
2	GB	9	BB
3	FB	10	K
4	FB	11	2B
5	3B	12	GB
6	GB	13	1B
7	1B	14	FB

FIGURE 1-3 Realignment of Bill Terry replica ASB disk into On-Base (light) and Out (heavy) mega-slices.

If we just consider all of the results split between just these two categories, we have reduced our multinomial distribution (a distribution over many categories) into a *binomial distribution* (a distribution over exactly two categories). The probability of occurrence for each of these two events is just the sum of the probabilities for each of the results contained within the individual event categories. For example, the probability of getting on base in a plate appearance is simply the sum of the individual probabilities for the single, double, triple, home-run, and base-on-balls results. Table 1-2 presents the final numerical results of this categorical distillation for Bill Terry in ASB.

We can see this intuitively by examining a player disk. In ASB, the batter gets on base whenever the spinner stops in a single, double, triple, home-run, or base-on-balls slice. We could reposition all of these slices so that they are next to each other. We would then have one segment of the disk with On-Base results and one with Out results. The block of On-Base slices contains an arc that is the sum of the arcs for each of the single, double, triple, home-run, or base-on-balls slices. Thus, on any single spin, the probability of obtaining an On-Base result is the sum of the probabilities of the individual results which compose the block. Figure 1-3 demonstrates how Bill Terry's disk in Figure 1-2 can be re-arranged to create the On-Base and Out mega-slices.

Model Assumptions of All-Star Baseball

ASB is a very simple model of baseball. It is accessible as a game for children of elementary school age because it simulates only the *hitting* ability of different players. Let us list some of the assumptions made by the model:

- Defensive fielding ability has no effect on the outcome of a plate appearance.
- A hitter's ability *never* changes.
- The ball park has no effect on the game.
- Weather conditions have no effect on the game.
- All players have the same ability to steal and run the bases.
- All hitters have the same ability to bunt and execute the hit and run.
- The pitcher has no effect on the outcome of a plate appearance.

This last assumption had a very curious effect on managerial strategy in the game. Versions of ASB in the 1950s included batting disks for pitchers. Because the ability of the pitcher to get batters out was not modeled in the game, managers chose pitchers purely on their ability to hit. (Don Newcombe of the Brooklyn Dodgers, a lifetime .271 hitter, was a particular favorite of ours. If Babe Ruth had been listed as a pitcher as well as an outfielder, there would be no question who would pitch the entire game without relief.) Of course, since the game came only with the best players from the past and present (1950s present, that is), the manager could also opt for the strategy of replacing the pitcher in every plate appearance. Why let the pitcher bat when you had second-tier stars like Jackie Jensen or Minnie Minoso just sitting on the bench? Actually, if one thought of the board game as an actual All-Star Game, this last strategy was the ultimate in realism.

The inclusion of pitchers in All-Star Baseball took a bizarre turn with the coming of the designated hitter rule. Pitcher disks were still part of the package, but were blank except for a photograph. As a critic of the DH rule and its elimination of that wonderful rarity, a clutch hit by the pitcher, at least one of us felt cheated by this turn of events in ASB. The pitcher's spot was always taken by a DH. Replacing information (the pitcher's hitting ability) with a colorful graphic photograph was an unfortunate sign of the times, foreshadowing the arrival of computer sports games, which hide their models behind eye-catching graphics.

All-Star Baseball was developed in 1941 by Ethan Allen, a former player. It lasted for decades, but now is out of print, having finally succumbed to the problem of not being realistic enough for adults and not exciting enough for today's generation of children.

The APBA Model: Introducing the Pitcher

The most obvious deficiency of All-Star Baseball as a model was the absence of any effect from pitching on outcomes at the plate. Successors to ASB provided a wide variety of innovations to baseball simulation, but generally they can be distinguished by the various ways in which they integrated pitching into their models. In looking at some of these other games, we will focus on this aspect of their models.

APBA Baseball, released in 1951 and still going strong, was the first tabletop game to take the effect of pitching into account. Each player (including pitchers) has a card simulating his batting record. Results are generated by tossing two six-sided dice. Each card has 36 equally probable results as determined by the toss of the dice. Basically, the system is the same as ASB, but with ASB disks replaced by cards and the spinner replaced by two dice.[2]

However, APBA Baseball introduced a level of complexity not found in ASB. Where ASB had a single table of reference to determine the play outcome, APBA Baseball has eight charts, one for each of the eight possible on-base situations:

- Nobody on.
- Man on first.
- Men on first and second.
- Men on first and third.
- Man on second.
- Men on second and third.
- Man on third.
- Bases loaded.

Within each chart, the outcome could be altered depending on the defensive ability of the team in the field and the ability of the pitcher. The abilities of pitchers are represented by different grades: A&B (best), A&C, A, B, C, or D. (APBA later upgraded its system in a "Master Game" version; it follows a similar scheme, but with expanded pitcher grades. The analysis described here

---

2 More information on APBA Baseball, as well as an errata sheet and other information about this book, can be found at our publisher's website, www.copernicusbooks.com.

refers to the earlier version of the game.) Depending on the base situation, the pitcher grade can change certain outcomes of a single into an out. Additional ratings were given to pitchers for their propensity to strike out or walk batters. The W (walk) rating could be particularly devastating; with runners on first and second, it changed a double play into a base-on-balls. We will focus only on the pitcher grades here.

Table 1-3 provides examples of pitchers with their APBA grades and key statistics for the season rated. At the top of the list is Joe Wood, who in 1912 had one of the greatest pitching seasons in baseball history. In recognition of this, APBA awarded him A&B—its highest possible pitcher rating. That year, Wood led his Boston Red Sox to a dramatic eight-game World Series victory, with one game ending in a tie when it was called because of darkness after 11 innings, and a Boston victory in the final game in extra innings with Wood appearing in relief! (Wood deserves more attention than he gets these days: after a brief but spectacular pitching career, his smoking fastball deserted him, and in 1915 he converted to an outfielder. He had always been a good-hitting pitcher; he had a .290 batting average to supplement his pitching in the golden year of 1912. He was an outfielder for the 1920 World Champion Cleveland Indians, and at the end of his career, from 1932 to 1942, was a baseball coach at Yale University.)

Jay Hannah "Dizzy" Dean's 1934 season was almost a match for Wood's amazing 1912, so APBA gave Dean its second highest grade. Like Wood, he also led his team, the St. Louis Cardinals, to a world championship, winning the final (seventh) game—an 11–0 shutout of the Detroit Tigers. In a continuing parallel, Dean also had a career of comet-like brilliance and brevity. Unfortunately, he did not have Wood's hitting ability as a fallback, but he did have a down-home folksy way of expressing himself that led to a long career as a broadcaster. "I never keep a scorecard or the batting averages. I hate statics. What I got to know I keep in my haid." (Voices of Baseball, p. 182).

Jim Palmer, like Dean, has continued his association with baseball through broadcasting. His Cy Young season for the Baltimore Orioles, in 1973, brought him an A grade from APBA. The season was so outstanding for Palmer that he also placed second (to Reggie Jackson) in the balloting for Most Valuable Player in the American League.

We couldn't write a book about baseball without including one of our boyhood heroes, Robin Roberts. In 1950, Roberts had his first of six consecutive 20-win seasons. He pitched (and won) a complete game in extra innings against the Brooklyn Dodgers to capture the pennant for the Phillies that year. He started

Grade	Pitcher	Season	ERA	Wins	Losses
A&B	Joe Wood	1912	1.91	34	5
A&C	Dizzy Dean	1934	2.65	30	7
A	Jim Palmer	1973	2.40	22	9
B	Robin Roberts	1950	3.02	20	11
C	Allie Reynolds	1950	3.73	16	12
D	John Odom	1973	4.50	5	12

TABLE 1-3   Examples of Pitcher Grades in APBA Baseball

only one game in the World Series, losing 2–1 in extra innings when he gave up a home run to Joe Dimaggio. Roberts' season was awarded a B grade by APBA.

The pitcher who bested Roberts in that 2–1 World Series game was Allie Reynolds, who never had a losing record from 1947 through 1954 with the New York Yankees. However, because he was closer to an average pitcher in 1950 than in other years, APBA rated him a Grade C pitcher.

Finally, John "Blue Moon" Odom was the weakest starting pitcher on a great pitching staff for the 1973 World Champion Oakland A's. Odom had several fine seasons, but 1973 signaled the end of his career. Odom was given a D rating for his pitching performance that year.

Table 1-3 and these brief vignettes give some feel for the pitching grades in APBA Baseball. A&B and A&C are the highest grades, reserved only for the greatest seasonal pitching performances in baseball history. Grade A pitchers are definite Cy Young Award candidates and often winners. Grade B pitchers had very good seasons, perhaps the number-2 pitcher on a very good staff or the ace on a weaker staff. Grade C pitchers are competent average starters. Grade D pitchers are the weakest performers.

Table 1-4 summarizes the conversion of results from the pitcher grades. The only numbers affected by the pitcher's grade on play charts in APBA Baseball are results #7, #8, #9, and #10. Each cell of Table 1-4 presents the conversion of these four results in sequence (#7, #8, #9, #10) for each pitcher grade in each base situation. S represents a single while O represents an out. For example, for a Grade B pitcher with a runner on second base, #7 and #9 result in a single while #8 and #10 result in an out. This table is somewhat of a simplification; the out results with runners on third base often result in sacrifice flies, for example. The overall effect on batting is not easy to estimate because it changes for different base situations.

Pitcher Grade	Empty	First	Second	Third	First & Second	First & Third	Second & Third	Full
A&B	SOOS	**OOSS**	SOOO	OOOS	SOOO	OOOO	OOOO	SOOS
A&C	SOOS	**OOOS**	SOOO	OSOS	SOOO	SOOO	OSOO	SOOS
A	SOOS	OOSS	SOOO	OSOS	SOOO	SOOO	OSOO	SOOS
B	SOSS	SOSS	SOSO	SOSS	SOSO	OSSS	SOOS	SOSS
C	SSOS	SSOS	SSOS	SSOS	SSOS	SOSS	SSOO	SSOS
D	SSSS	SSSS	SSSS	SSSS	SSSS	SSSS	SSSS	SSSS

TABLE 1-4   Effects of Pitcher Grade on Play Results in APBA Baseball

We expect that pitchers with better ratings would produce more outs than those with lower ratings. Referring again to Table 1-4, a quick sum of Os across the rows produces 23 for A&B, 21 for A&C, 20 for A, 11 for B, 9 for C, and 0 for D. This provides some support for the ordering of pitcher abilities found in the table. Another test is for internal consistency within each column. An examination of the columns of the table indicates that generally they are consistent: given any cell, the number of out results is equal to or greater than the number of out results in the cell below it. The one exception is noted by the two cells in boldface. With a runner on first base, an A&C pitcher is *better* than an A&B pitcher. Since we expect a better pitcher to produce better results in every base situation, this is a troubling inconsistency (corrected later in the "Master Game" version).

To get a better feel for the effects of pitcher grades, we have to consider the frequency of results #7 through #10 on player cards. APBA Baseball was introduced with cards replicating the 1950 season, the year of the first Philadelphia Phillies pennant in 35 years. In honor of this affectionately remembered team, we have chosen three of these Whiz Kids to examine the pitcher effects. Table 1-5 shows the frequency of these results for Del Ennis (a slugging left fielder not so different from Greg Luzinski), Andy Seminick (a tough catcher, a 1950s version of Darren Daulton), and Willie Jones (predecessor to Dick Allen, Mike Schmidt, and Scott Rolen at third base). We have included their nicknames from their APBA cards, a charming feature not found in other games.

For each player, we can calculate the effect of pitching grades within each base situation. We will use pitching grade D as our standard and calculate the amount subtracted from each player's Pr(On Base).[3] Here are two examples of this calculation for Del Ennis:

Player	#7	#8	#9	#10
Del "Skinny" Ennis	2	4	2	0
Willie "Puddin' Head" Jones	2	2	2	1
Andy Seminick	2	3	2	0

TABLE 1-5   Frequency of Play Results Affected by Pitcher Grade in APBA Baseball for Three Members of the 1950 Phillies (the Whiz Kids)

- *Del Ennis is batting against an A&B pitcher with runners on first and third bases.* According to Table 1-4, results #7 through #10 all produce outs. Of the 36 results on Ennis's card, 8 produce results #7 through #10 (see Table 1-5 above). So, Ennis has 8/36 = .222 less chance of getting on base against an A&B pitcher than he does against a D pitcher. That is,

$$\text{Pr(On Base vs. A\&B)} = \text{Pr(On Base vs. D)} - .222$$

when runners are on first and third. Note that this is the maximum possible effect, since the A&B pitcher has changed every single that could possibly be affected into an out.

- *Del Ennis is batting against a B pitcher with runners on second and third bases.* According to Table 1-4, results #8 and #9 produce outs. Of the 36 results on Ennis's card, 6 produce results #8 and #9 (see Table 1-5 above). So, Ennis has 6/36 = .167 less chance of getting on base versus a B pitcher as compared to a D pitcher. That is,

$$\text{Pr(On Base vs. B)} = \text{Pr(On Base vs. D)} - .167$$

when runners are on second and third.

Table 1-6 shows the effect of each pitcher grade on Skinny's Pr(On Base) in each base situation. Each cell shows the result of a calculation like the ones just described. That is, the value in each cell is the difference in Pr(On Base) between the relevant pitching grade and a Grade D pitcher. In general, we see that a Grade C pitcher reduces Pr(On Base) about .06, a Grade B pitcher reduces the

---

[3] This notation will be used throughout the book. Pr(Event) is the probability that Event takes place. For example, here it refers to the probability of getting on base.

Pitcher Grade	Empty	First	Second	Third	First and Second	First and Third	Second and Third	Full
A&B	.167	**.167**	.167	.222	.167	.222	.222	.167
A&C	.167	**.222**	.167	.111	.167	.167	**.111**	.167
A	.167	.167	.167	.111	.167	.167	**.111**	.167
B	.111	.111	.111	.111	.111	**.056**	**.167**	.111
C	.056	.056	.056	.056	.056	**.111**	.056	.056
D	.000	.000	.000	.000	.000	.000	.000	.000

TABLE 1-6   Reduction in Del Ennis's Pr(On Base) in Different Base Situations When Opposed by a Pitcher with Grade D or Better

probability about .11, and a Grade A pitcher reduces it about .16. A Grade A&C pitcher shows very little improvement over an A pitcher, but a Grade A&B pitcher demonstrates a great improvement in most base situations with a runner on third base.

The boldface entries in the runner-on-first ("First") column provides a numerical value for the inconsistency described earlier and highlighted in the shaded cells in Table 1-4. However, the use of actual values for the frequencies of the #7 through #10 results has unearthed several other inconsistencies, also identified by boldface entries in Table 1-6. With runners on first and third, Ennis has a better chance of getting on base versus a Grade B pitcher than against a Grade C pitcher. However, if we move that runner on first to second base, the ability of the Grade B pitcher improves enormously, to such a degree that he is a greater adversary than a Grade A pitcher or even a Grade A&C pitcher.

Unlike the "Runner on First" inconsistency, which holds for all batters with a #9 result, the last two inconsistencies are dependent on the distribution of frequencies of results #7 through #10. If we use the distribution for Willie Jones, these last two inconsistencies disappear. And then it reappears with Andy Seminick. However, for both of these players, the maximum effect is reduced from .22 to .19 = 7/36.

In many ways, this model is very similar to ASB's, in that it is only hitting that is modeled in great detail. Pitching (and through a similar mechanism, fielding) influences only selected results. The frequency of extra-base hits was not affected by pitcher grades. Pitchers were given a separate indicator if they gave up more walks than average; this rating turned certain out results into walks.

It should be noted that while the implementation of a pitcher effect was a major development in APBA Baseball, the use of two dice in place of a spinner was not necessarily an improvement. Dice and a spinner are equivalent randomizing devices from a theoretical viewpoint (that is, apart from trying to influence the spin or using loaded dice). But they are very different in at least one important respect. The spinner is an analog randomizing device which can assume *any* value (direction) on a continuum between 0 and 360 degrees. Dice restrict randomization to a finite set of results with a minimum probability that can be represented. This presents certain problems.

Consider a batter with 7 triples in 500 plate appearances. The proportion of triples is 7/500 or .014. In ASB, we can easily replicate this performance on the disk by assigning a pie slice with an arc of 5 degrees (1.4 percent of 360 degrees). In APBA Baseball, on the other hand, the minimum we could assign to a triple would be one of the 36 paired results on a batting card, or 1/36 = .028. This is almost twice the probability we wish to be represented! Since the minimum assignment we can make is 1 of the 36 results, we are stuck with an imprecise simulation of the batter's ability.

This shortcoming in the APBA model may not seem terribly important until one realizes that the probability of any event in any situation is restricted to multiples of this minimum amount, or quantum, of 0.28. That is, in any situation for any batter an event's probability can only be one of the following values: .028 (1 out of 36 results), .056 (2 out of 36 results), .084 (3 out of 36 results), and so forth. It is true that on many APBA cards certain dice rolls produce a #0 result, which requires a second roll of the dice with the results taken from a second column on the card. While we don't get into analyzing these cards in this book, we do admit that they allow for a finer gradation of probabilities, especially for relatively rare events. And we do acknowledge that APBA's use of dice rather than a spinner does avoid certain problems—like spins that end up on the line between slices. Nonetheless, the point here is that though they may have certain advantages over a spinner, dice are inherently less precise in representing performance.

Strat-O-Matic Baseball: The Independent Model

The development of *Strat-O-Matic (SOM) Baseball* in 1962 provided a new variation on the pitcher's effect. Like APBA Baseball, this game uses dice as a randomizing device, and each everyday player has a separate card that simulates his hitting ability. In an inversion of APBA Baseball, however, each pitcher has a card which simulates *pitching ability*. (A separate numerical rating is given for each pitcher's hitting ability.) SOM Baseball was the first game to give pitching this level of modeling detail.

The mechanism for the game is simple. To initiate a play, three six-sided dice (one white and two red) are rolled. The white die is used to determine the card column for the result. Batters' cards have columns numbered 1, 2, and 3, while pitchers' cards have columns 4, 5, and 6. The two red dice are summed, and the value is used to find the resulting play within the relevant column. On occasion, an extra randomizing device, a so-called "split deck" of cards numbered 1 through 20, is needed when multiple play results are listed for a dice roll. This is similar to the extra column on APBA cards. (Apparently it is used to increase the precision of the simulation.)

The important element on which to focus is the even split of results between the batter and the pitcher. Upon closer inspection of pitcher cards, one finds that the split is actually 50–50 between the batter and the *defense* (pitching and fielding). On each pitcher's card, 28 percent of the resulting plays require an extra randomization (using the ever-present split deck), which references the ability of a specified fielder to make a great play or avoid making an error.

We can summarize the ability of a batter versus a pitcher on the defensive team as follows:

50% Batter Ability + 50% Team Defensive Ability

In the SOM model, Team Defensive Ability is 28 percent fielding and 72 percent pitching. So the ability of a batter versus a pitcher on the defensive team is actually:

50% Batter Ability + 50% × (72% Pitcher Ability + 28% Team Fielding Ability)

or

50% Batter Ability + 36% Pitcher Ability + 14% Team Fielding Ability.

An old saying goes that pitching is 75 percent of winning in Major League Baseball, but it's less than half that in Strat-O-Matic Baseball.

Let's examine some of the players we investigated in APBA Baseball. Checking out a Del Ennis card for the 1950 season we find that 41.7 percent of the results put Del on base. Allie Reynolds, a Grade C pitcher in APBA Baseball, faced Ennis in the 1950 World Series. In Strat-O-Matic Baseball, Reynolds would put a batter on base in 40.1 percent of the results he controls. Apart from the Fielding Ability of the 1950 Yankees, Ennis has about as much chance of getting on base from his own batting ability as from Reynolds' pitching ability. On the other hand, Vic Raschi, a Grade B pitcher in APBA Baseball, has only a 33.7 percent chance of putting the batter on base in the results he controls. In this case, Ennis is much better off if the white die places the result on his batting card. The difference in pitching ability between Raschi and Reynolds can be summarized as follows:

$$40.1\% - 33.7\% = 6.4\%$$

That is, of the results controlled by pitchers, Raschi puts batters on base 6.4 percent less than Reynolds.

However, to get the true effect, we must also account for the frequency that the Pitcher Ability is used (36 percent of the time). So, the overall effect of using Raschi instead of Reynolds is 36 percent of 6.4 percent, which equals 2.3 percent. In terms of probability, Raschi subtracts .023 from the probability of getting on base when compared to Reynolds. Note that this is less than half the effect modeled between these two pitchers in APBA Baseball when facing Del Ennis.

This discussion has focused only on pitching. However, for the sake of completeness, we will make a brief foray into fielding, mainly because the SOM model allows us to do this with little added complexity. Fielders are given ratings not unlike pitching grades in APBA Baseball. The ratings range from 1 (the best) to 4 (the worst). As noted earlier, 28 percent of the results on the pitcher's card require a new random result based on the rating of a particular fielder. For example, one such result on a pitcher's card references the Fielding Ability of the left fielder. Another random number (derived from the split deck) is looked up on a fielding chart under the column for the Fielding Ability of the left fielder. A left fielder with a 1 rating cannot give a batter a hit while a left fielder with a 4 rating gives the batter a hit 30 percent of the time.

To get a feel for the range of differences in fielding, we will look at a team with the best possible rating (1) at each position versus the worst possible fielding team, with a rating of 4 at each position. With a 1 rating at each position, fielders would give up no extra hits to batters; in terms of the batting formula above, Fielding Ability is .000. On the other hand (and it makes us cringe to think about it), if the fielding team had a 4 rating at each position, on average the Fielding Ability of the team would be .324. That is, when the Fielding Ability of the worst fielding team is referenced, the batter gets a hit about 32 percent of the time. Since fielding is referenced in 14 percent of the plate appearances, the maximum effect from fielding is 14 percent of .323, or .045.

The interesting feature of SOM Baseball's hitting model is that it is purely *additive*. To find the probability of a batter getting on base, we need only add the abilities of the batter, pitcher, and fielders. Consider Del Ennis in 1950. His Batter Ability is .417. Since this is used half the time, the batter contribution to the probability of getting on base is .417/2 = .208. If he is facing Allie Reynolds supported by the best possible fielding team, then his probability of getting on base is just the sum of their contributions:

$$.208 + .144 + .000 = .352$$

Replacing Reynolds with Raschi decreases the probability:

$$.208 + .121 + .000 = .329$$

If we replace the best fielding team with the worst possible fielding team, both probabilities increase. For Reynolds:

$$.208 + .144 + .045 = .397$$

and for Raschi:

$$.208 + .121 + .045 = .374$$

(Note that for simplicity we are ignoring that the fielding rating for the team is influenced by the fielding ratings for Raschi and Reynolds themselves, a minor effect.)

Figure 1-4 plots a batter's probability of getting on base as his ability increases. The dashed line presents this probability for All-Star Baseball; this is a 45-degree diagonal, or a line with slope 1, since the x and y values are always equal: the probability of the batter getting on base is identical to the batter's ability in getting on base as described on the ASB disk. However, in Strat-O-

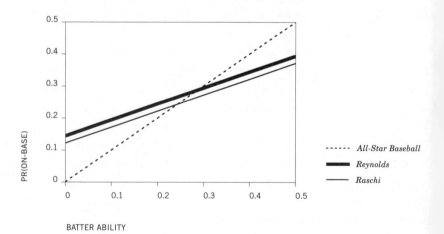

FIGURE 1-4   Plots of pitcher effects in Strat-O-Matic Baseball and All-Star Baseball.

Matic Baseball, the pitcher and fielders, as well as the batter, influence this probability. For this plot, we have assumed the best fielding team, which adds 0 to the probability. Two lines are shown, one for Allie Reynolds and one for Vic Raschi, in SOM Baseball. We see that both lines have identical slopes, .5, which is the result of the SOM batter influence described earlier. The difference in their lines resides in their intercepts, or starting point: the point at which the line crosses the y axis. This difference is .023, as described earlier. In fact, if we plotted a line for any pitcher in SOM Baseball, it would have the same slope (.5) but a different intercept. Thus, the relative skills of pitchers are represented by shifting the line up or down.

The SOM model, in terms of consistency, is an improvement over the APBA model. The pitcher with better ability will tend to be better in all situations. The additivity described above guarantees this feature. On the other hand, something is lost as well: the interaction between pitcher and batter is not taken into account. In APBA Baseball, the effect of the pitcher depended on how his characteristics (defined by the pitching grade) interacted with those of the batter (defined by his batting card). The two features worked together to produce a result. In Strat-O-Matic Baseball, these characteristics do not interact; they are completely independent. The batter has his ability, and the pitcher has his; a certain percentage of the time one is used and the other is ignored. The two are not fused to produce a result.

At times this additivity can produce strange effects, particularly with respect to the frequency of triples and home runs. In SOM Baseball, Robin Roberts' 1950 pitching ability is expected to add about 9 home runs and 4 triples for every 1,000 batters faced. This addition is the same whether Roberts faces Joe Dimaggio, Stan Musial, Whitey Ford, Ozzie Smith, Mark McGwire, or Randy Johnson. All batters have an equal chance at obtaining these home runs and triples. However, more so than other events, home runs are mostly dependent on the power of the batter, while triples are mostly dependent on his base-running speed.[4] Other tabletop games have taken the basic Strat-O-Matic model and added their own wrinkles to it. *Pursue the Pennant* adopted the SOM model and added more detailed results plus other effects. In *Ball Park Baseball*, the batter can have exceptions that overrule the result on a pitcher's card and vice versa.

---

[4] Advanced versions of Strat-O-Matic Baseball make gross adjustments for the HR problem by designating some hitters as capable of obtaining HRs on the pitcher's card while others do not. Essentially, this introduces an interaction effect into the basic SOM model.

Sports Illustrated Baseball: The Interactive Model

Sports Illustrated (SI) Baseball was developed in 1972 by David S. Neft, co-author of *The Sports Encyclopedia: Baseball*. This tabletop game used a two-tier or hierarchical approach to produce an interactive model for hitting in place of SOM's independent model. In SI Baseball, the manager of the team in the field rolls a set of three dice to obtain a result on the pitcher's chart. The result could be a strikeout, a walk, a single, or "Batter Swings." This last result, the most common occurrence, allows the player managing the team at bat to roll the dice to obtain a play on the batter's chart.

Each plate appearance is a two-step process. There is a hierarchy of precedence in the structure of the plate appearance. The pitcher controls the outcome of the plate appearance until his ability is used to finalize the result or relinquish control to the batter. Basically, the batter must get past the pitcher before he is able to use his batting capabilities. Given the nature of baseball, in which the pitcher does control the game process, this model has some intuitive appeal.[5]

Let's see how this model looks from a probability perspective. The probability of the batter getting on base *in the first step* under the pitcher's control is calculated as follows:

$$Pr(\text{On Base in Step 1}) = Pr(\text{On Base on Pitcher Chart})$$

Similarly, we can calculate the probability of the batter getting on base *in the second step* using his own batting capability:

$$Pr(\text{On Base in Step 2}) = Pr(\text{On Base on Hitter Chart})$$

However, in order to obtain the overall probability of a batter getting on base we must combine the probabilities from the two steps. The first step (attempting to get on base via the pitcher's chart) is always used, but the second step (attempting to get on base via the hitter's chart) is only used when the "Batter Swings" result is obtained on the pitcher's chart. Thus, the probability of a batter getting on base is:

$$Pr(\text{On Base on Pitcher Chart}) +$$
$$Pr(\text{Batter Swings on Pitcher Chart}) \times Pr(\text{On Base on Hitter Chart})$$

---

[5] Interestingly, Kevin Hastings independently developed a modification of the Strat-O-Matic model with interactive effects. A close examination of the model presented in his Winter 1999 *Chance* paper shows it to be basically equivalent to the one used in Sports Illustrated Baseball.

Pitcher	Pr(On-Base on Pitcher Chart)	Pr(Batter Swings on Pitcher Chart)	Pr(On-Base)
Robin Roberts	0.033	0.767	.033 + .767 × Pr(On-Base on Hitter Chart)
Bobby Shantz	0.057	0.905	.057 + .905 × Pr(On-Base on Hitter Chart)
Vida Blue	0.071	0.638	.071 + .638 × Pr(On-Base on Hitter Chart)

TABLE 1-7   Examples of SI Baseball Pitcher Effects

The multiplication of probabilities from the Pitcher Chart and from the Hitter Chart produce an interaction between batter and pitcher. This is different from the Strat-O-Matic hitting model, which only adds the effects from the pitcher and hitter.

Table 1-7 shows the pitcher effects for three pitchers from SI Baseball. The first column presents the probability that the pitcher puts the batter on base automatically, without any reference to the batter's skills. The second column presents the probability that the "Batter Swings," requiring a reference to the batter's hitting skills. The third column presents (as a formula) the way these two values are combined by the SI Baseball model to calculate the probability of a batter getting on base given knowledge of his hitting skills. Figure 1-5 presents a plot of this formula for each pitcher. Note that the pitcher effects are all straight lines, but with different slopes and starting points (intercepts).

One of the pitchers is our old friend Robin Roberts. The effect shown here is taken from his career record as it was captured in SI Baseball's set of All-Time All-Star Teams for the sixteen original franchises. Roberts made the squad for the Philadelphia Phillies. As the teams were selected in 1972, the Phillies are missing many of the franchise's finest players from the team's golden era in the late 70's and early 80's. (There's no Mike Schmidt or Steve Carlton.) However, it's safe to say that Roberts, a Hall of Famer, would still make the squad if it were picked today. Roberts was well-known for his control. This aspect of his pitching skill is reflected in the low value for automatic walks and the low intercept of his line.

Bobby Shantz is not well known today, but in 1952, he and Roberts may have produced the greatest starting pitcher tandem from the same city. Roberts went 28–7 for the Phillies while Shantz went 24–7 for the Philadelphia Athletics. Both would undoubtedly have won the Cy Young Award, but the award's creation was still four years in the future (and wasn't given in both leagues separately for another 11 years). While Roberts finished second to Hank Sauer, a

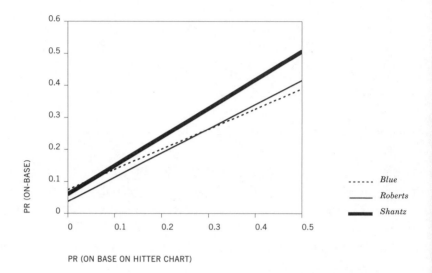

PR (ON BASE ON HITTER CHART)

FIGURE   **1-5** Plot of pitcher effects in SI Baseball from Table 1-7.

Cubs slugger, in the National League MVP balloting, Shantz did win the American League award, beating out another old friend, Allie Reynolds. While Shantz never had another year like 1952, he remained in baseball for 12 more years, primarily as a relief pitcher. He was also noted as a fielder, winning eight consecutive Gold Gloves. (He probably would have won more, but the award was not created until 1957, his ninth year in the major leagues.) Overall, Shantz's record was good enough to place him on the list of SI Baseball's All-Time All-Star Athletics in 1972. Like Roberts, the effect shown here represents his career record.

Shantz and Roberts provide an interesting comparison. Unlike Roberts, for Shantz the overall probability of the batter getting on base is *greater* than the probability of his getting on base from the hitter's chart in almost all reasonable cases. In Figure 1-5, we see that Roberts' performance completely dominates Shantz's performance; every batter has less chance of getting on base opposing Roberts than opposing Shantz. Indeed, the better the hitter, the bigger the difference between the effects of Roberts and Shantz. We can see this by the ever-widening gap between their lines in Figure 1-5, as hitter skill (represented by the probability of getting on base from the hitter chart) increases. One can interpret this as saying that good pitching becomes more important as the skill of the batter faced increases. Or, good hitters feed off of poorer pitching. This is what we mean when we say that the pitcher and batter interact.

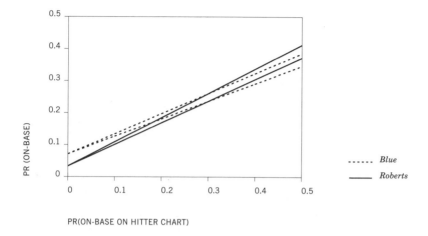

PR(ON-BASE ON HITTER CHART)

FIGURE 1-6   The effects of fielding in SI Baseball.

The third pitcher, Vida Blue, shares some similarities with Shantz. Both had their best years with the Athletics. Blue had a more significant career, with more wins and major contributions to several World Championship teams, but like Shantz he is best remembered for that one great season. In 1971, Blue went 24–8. He became the first Athletic to win the Cy Young Award and the first Athletic (since Shantz) to win the MVP Award. The pitcher effect in the figure represents his performance in 1971. From the plot we see that Blue had the least control of the three. Against weaker hitters (e.g. pitchers) with probabilities of getting on base between .1 and .2, he is better than Shantz but worse than Roberts. However, against better hitters with on-base probabilities greater than .3, his ability to get batters out exceeds his control problem and makes him a more difficult adversary than Roberts.

Table 1-7 and Figure 1-5 were created without consideration of the fielding team's abilities. SI Baseball integrates fielding into the pitcher's chart. Basically, teams with better fielders decrease the Pr(Batter Swings on Pitch Chart). This decrease ranges from .010 to .088. Even with the best fielding team behind him and the worst behind Roberts, Roberts still dominates Shantz's performance. We can see this simply by subtracting .088 from Shantz's Pr(Batter Swings on Pitcher Chart) = .905 − .088 = .817, which is still greater than .767, Roberts' basic value for this probability. The issue is not so clear between Roberts and Blue. Figure 1-6 shows the range of fielding effects on Pr(On Base) for Roberts and Blue. The solid lines identify the pitcher effects for the worst fielding team (high line) and the best fielding team (low line) for Roberts. The dashed lines perform the same function for Blue. We see that, for better hitters, the fielding team behind these pitchers can make a difference as to which is better overall.

Which Model Is Best?

It is difficult to say which system is best. We both have owned, played, and still enjoy all of these tabletop games. Much of the model design depended on their being played for entertainment, and not exclusively, as here, on how they stand up to analysis. Additional complexity may add to the realism of the model but detract from its playability. In some ways, this is in the true spirit of the construction of statistical models, where added complexity is not included unless it produces a significant improvement in its picture of reality. Unlike their computer counterparts, the models used in tabletop games are exposed to the player. Instead of locking the model inside compiled code, with no one able to review how results are generated, the developers and publishers of these games have left themselves bravely open to criticism.

We have reviewed these systems at the most basic level, the probability of getting on base. Each game also goes into varying detail on the types of events affected by the pitcher. Strat-O-Matic Baseball pitcher cards vary the distribution of types of hits depending on the pitcher's record; thus, pitchers who give up a lot of home runs (like Robin Roberts) are represented by cards with a greater probability of home runs. SI Baseball does not do this. For the most part, pitchers with the same "Batter Swings" probability have the same effect on all hitters, proportionately decreasing the probability of each type of hit.

Many of these games have evolved over the years. They have integrated new effects such as righty/lefty batting effects, ballpark effects, and performance in critical situations (clutch effects). Strat-O-Matic Baseball has moved from a basic version to an advanced version to a super advanced version. APBA Baseball has created a "Master Game." For simplicity of exposition, all analysis here has used the most basic version of each game.

This is not a real impediment to our discussion, however, because the central point has not been the details, but the general nature of the models, the distinction between models with no pitcher effect (All-Star Baseball), models with additive pitcher effects (Strat-O-Matic Baseball), and models with interactive pitcher effects (APBA Baseball and SI Baseball). None of these systems is perfect, but then what model is? Models attempt to capture reality to an extent limited by the needs of their users and the data available to support their validity.

Each model has its strengths and weaknesses. All-Star Baseball has the least sophisticated model, but its simplicity allows it to be played quickly and to introduce younger children to baseball rules, history, and even some important statistical concepts (fractions and pie charts). At the other end of the scale, Strat-O-Matic Baseball puts the greatest effort into capturing the details of the distribution of plays for batters and pitchers. However, doing this with a game system which attempts to derive results from one roll of three dice produces a model which at times can overestimate the probability of rare events (home runs and triples by weaker batters). APBA Baseball and SI Baseball occupy the middle ground. They sacrifice much of the detail in pitcher effects included in SOM Baseball, but have a more interesting interactive model between the pitcher and batter. SI Baseball provides more detail than APBA but at a cost in terms of playability; it almost always requires two dice rolls and chart references to obtain a play result.

We will complete the chapter on a note of harmony. Instead of focusing on differences, let's examine similarities. All of these games model baseball as series of events randomly generated from player characteristics. The games do not model any sort of momentum effect. A pitcher that gets rocked for a home run has the same chance of getting the next batter out; his ability is unaffected by the previous unsettling event. Player abilities are fixed; they have good days and bad days only as a result of the random variability in the twirl of the spinner or the roll of the dice.

Another common feature of these games is the absolutely precise information that tabletop game managers have about each player. As we have shown, proper analysis of player cards and disks allows each tabletop manager to know *exactly* each player's ability to perform. In fact, this may be the least realistic aspect of these games as models, because real managers are limited to observations and measurements of player performance that are the product of ability and chance. In subsequent chapters, we will discuss some of the implications of chance and random variability and how we can explore measurements of baseball performance through the fog of chance and identify true ability and significant effects in the play of the game.

EXPLORING Baseball Data

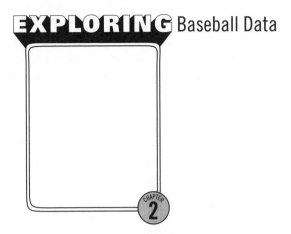

CHAPTER 2

Baseball data consist of a large number of counts and averages that are supposedly helpful in evaluating players and describing the game. The baseball fan is bombarded, inundated, overwhelmed with statistics. It may be that it is difficult for the fan to understand the relevance of a particular baseball statistic for the simple reason that there are so many of them competing for his or her attention. Is it slugging percentage that's really important, or on-base percentage?

One job of the professional statistician is to organize and summarize data effectively. But what does "effectively" mean? First, we want to present the data with graphs or charts that make it easy to see general patterns in the numbers. Once we understand the basic patterns in data, we look for unusual data values (say a Mark McGwire slugging percentage or a Greg Maddux earned-run average) that appear to deviate from the general patterns. Taken together, all these methods of organizing and summarizing data are called *data analysis*.

Exploring Hitting Data

In this chapter, we introduce some basic tools of data analysis by exploring some hitting data, starting at the very beginning. A team wins a baseball game by scoring more runs than its opponent. How does a team score runs? Essentially it is a two-step process. First, batters get on base by getting hits or walks, by benefiting from opponents' errors, or by being hit by a pitched ball. Second, these runners are advanced to home by subsequent hits, walks, errors, and hit batters.

The best way of advancing runners is by means of a particular type of hit—a home run—which scores all of the players on base and the batter. Since it is important both to get on base and to advance runners, a typical team's lineup will consist of several types of hitters. The first and second hitter in a team's lineup are supposedly good at getting on base, and the batters in the middle of the lineup are typically powerful hitters who are good at advancing runners.

A Batch of On-Base Percentages

The standard measure for judging how good a batter is in getting on base is the On-Base Percentage, abbreviated OBP. Basically the OBP is the fraction of plate appearances where the batter gets on base. The Major League Baseball web site tells one how to compute OBP: divide the total number of Hits (H) plus Bases on Balls (BB) plus Hit by Pitch (HP) by the total number of At-Bats (AB) plus Bases on Balls (BB) plus Hit by Pitch (HP) plus Sacrifice Flies (SF). Using mathematical notation and the above abbreviations, the formula for OBP is:

$$OBP = \frac{H + BB + HP}{AB + BB + HP + SF}$$

Let's illustrate computing OBP for the 1999 Roberto Alomar. Table 2-1 displays Alomar's season statistics. We compute his on-base percentage in the following equation, which tells us that Alomar gets on base roughly 42 percent of the time:

$$OBP = \frac{182 + 99 + 7}{563 + 99 + 7 + 13} = .422$$

Now, is .422 a high OBP value? Is it one of the best values among American League players? How does it compare to "typical" hitters in the American League? We suspect that Alomar's value is high, since by reputation he is known as one of the best hitters in baseball. Common sense tells us he probably is very effective in getting on base. The question is, how can we use data to confirm (or not confirm) what common sense tells us?

	AB	H	BB	SH	SF	HP
Roberto Alomar	563	182	99	12	13	7

TABLE 2-1   1999 Season Batting Statistics for Roberto Alomar

The Major League Baseball website lists the OBP values for all 395 American League players who hit during the 1999 season. Looking over the list, we see many players who had small numbers of at-bats during the season. We don't want to compare Alomar with everyone—it would be inappropriate, for example, to compare him with a part-time player (say, a fielding specialist) who had only a few at-bats. It does seem reasonable, though, to compare his OBP value with those of AL players who played regularly during the 1999 season. We arbitrarily decide that a player is a "regular" if he had at least 400 plate appearances during the season. (Here the number of plate appearances is AB plus BB plus HP plus SF.) Using this definition of "regular," Table 2-2 shows the OBPs for the 108 regular American League players in 1999. This table of OBPs is hard to decipher. To better view these values, we will introduce a few simple graphical methods that statisticians find useful.

Simple Graphs

One basic method for organizing and displaying a small amount of data is a *stemplot*. This graph might appear odd at first glance, but it is a quick and effective way of organizing data.

Consider Alomar's OBP value, .422. Ignore the decimal point and break the value into two parts, which we call a *stem* and a *leaf*—in Alomar's case, the stem

0.330	0.353	0.379	0.427	0.267	0.352	0.358	0.356	0.325
0.422	0.335	0.360	0.310	0.351	0.353	0.343	0.372	0.414
0.336	0.369	0.304	0.339	0.307	0.312	0.353	0.329	0.426
0.404	0.280	0.366	0.418	0.309	0.311	0.391	0.355	0.315
0.365	0.333	0.377	0.422	0.280	0.405	0.349	0.305	0.33
0.387	0.361	0.328	0.331	0.338	0.341	0.420	0.328	0.358
0.330	0.346	0.373	0.378	0.414	0.361	0.339	0.366	0.39
0.331	0.335	0.354	0.324	0.438	0.447	0.362	0.393	0.397
0.400	0.307	0.346	0.384	0.340	0.405	0.341	0.344	0.343
0.337	0.365	0.287	0.405	0.413	0.363	0.442	0.371	0.435
0.334	0.302	0.308	0.358	0.315	0.327	0.363	0.347	0.315
0.307	0.358	0.336	0.384	0.393	0.357	0.357	0.387	0.354

TABLE 2-2   1999 On-Base Percentages for Regular American League Players

STEM LEAF
42 | 2

is 42 and the leaf is 2. (See Figure 2-1.) To draw a stemplot we first list all of the possible stems as a column and a vertical line to the right of the column, as shown in Figure 2-2. Then we record the OBP values by writing down only the leaf value on the right of the vertical line. For example, suppose we want to record the OBP values given in Table 2-3, where the stem and leaf for each are shown.

26
27
28
29
30
31
32
33
34
35
36
37
38
39
40
41
42
43
44

We record .330 (or 33 | 0) by writing a 0 on the 33 stem line, .353 (or 35 | 3) is recorded by writing a 3 next to the 35 stem line, and .379 (37 | 9) is recorded by writing 9 on the 37 stem line. Remember that each single digit on the right corresponds to one OBP value. So this stemplot:

$$28 \mid 007$$

corresponds to three players with the following OBP values: .280, .280, .287.

The stemplot in Figure 2-3 shows us the OBP values for all 108 American League regular players. It may be easier to see the pattern of OBPs by rotating the stemplot display 90 degrees so that the small OBPs are on the left. (See Figure 2-4.) This display tells us a lot about the pattern of OBPs for all American League regulars.

The first thing we notice in Figure 2-4 is the general *shape* of the group of OBPs. Most of the OBP values are in the .300–.390 range, and a relatively small number of hitters had OBP values smaller than .300 or higher than .400. So it is pretty common to have an OBP is the middle .300s and we should be somewhat impressed to see an OBP larger than .400 (like Alomar's) or depressed to see an OBP in the .200s.

OBP	STEM	LEAF
0.330	33	0
0.353	35	3
0.379	37	9

In Figure 2-5, we draw a smooth curve over the OBP values. This smooth mound-shaped curve is called a *normal curve*—it's a popular curve in statistics for representing a group of measurements. When they are recorded for

FIGURE 2-1   Breaking an OBP into a stem and a leaf.
FIGURE 2-2   List of all possible stem values from the data in Table 2-2.
TABLE 2-3   Some OBP Values with Corresponding Stem and Leaf Values

26	7
27	
28	007
29	
30	2457789
31	012555
32	457889
33	000113455667899
34	011346679
35	123334456778888
36	01123355669
37	123789
38	4477
39	01337
40	04555
41	2448
42	02267
43	58
44	27

large numbers of players, many baseball statistics—such as batting average, slugging percentage, or earned-run average—will result in a normal curve.

A related useful display of data is a *histogram*. To construct this picture of data, we group the OBPs into intervals of equal width, count the number of OBPs in each interval, and then make a bar graph where the height of the bar corresponds to the count in that interval. Suppose in this case that we decide to group the OBPs in the intervals .260–.269, .270–.279, .280–.289, and so on. Then we get a picture of the OBPs shown in Figure 2-6.

Note that this histogram gives us the same picture of the OBP data as we saw from the stemplot, since we are grouping the data in the same way. The histogram is perhaps a "prettier" display than the stemplot in Figure 2-3, but actually the stemplot is more informative, since we see the actual OBP values.

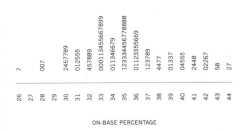

ON-BASE PERCENTAGE

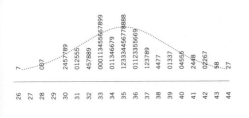

ON-BASE PERCENTAGE

Typical Values—the Mean and the Median

After we graph a group of baseball statistics as we've done above, we look for a *central* or *typical* value among all of the data. There are two popular ways of computing a typical value using averages: one average is called the *mean*, the other is the *median*.

The mean (or arithmetic average) is what you get when you add up all of the OBPs and

FIGURE 2-3   Stemplot of OBPs for 1999 American League regular players.
FIGURE 2-4   A stemplot that has been rotated 90 degrees.
FIGURE 2-5   Stemplot of OBP values with a smooth curve drawn on top.

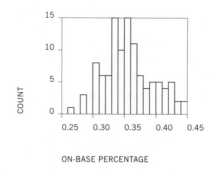

COUNT

ON-BASE PERCENTAGE

FIGURE 2-6   Histogram of OBPs for 1999 American League regular players.

divide by the number of data values. In the case of the data from Table 2-2, the computation of the mean would look like this:

$$Mean\ OBP = \frac{.330 + .353 + \ldots + .354}{108} = .356$$

The median is the middle value when all of the OBPs are arranged from smallest to largest—this number divides the data into a lower half and an upper half. The median of the data listed in Table 2-2 is .354.

Thus, the mean and median are both about .350, which tells us that a typical AL regular player will get on base about 35 percent of the time. Since the median is .354, we can say that half of the OBPs are smaller than .354, and half are larger.

Measures of Spread—Quartiles and the Standard Deviation

After we find a typical (mean and median) OBP for the data in Table 2-2, we want to say something about the *spread* of the OBPs. One simple way of describing the spread of a set of measurements uses the *lower* and *upper quartiles.* The quartiles divide the data into two extreme quarters—one-quarter of the data is smaller than the lower quartile, and another quarter of the data is greater than the upper quartile. Here the lower quartile of OBPs is .331, the upper quartile of OBPs is .382

So one-quarter of all the AL regulars have OBP values smaller than .331, and one-quarter have OBPs greater than .382. This means that half of all the American League OBPs are between .331 and .382. In Figure 2-7, we've redrawn our stemplot, showing the approximate location of the quartiles, and illustrating

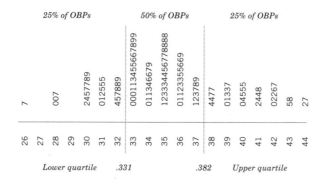

ON-BASE PERCENTAGE

FIGURE 2-7   Quartiles divide the OBPs into a lower quarter, a middle half, and an upper quarter.

how the lower and upper quartiles divide the group of OBPs.

Another measure of spread of a group of measurements is the *standard deviation*. This number gives the typical difference between an OBP value and its mean.[1] For our set of OBP numbers, the standard deviation is .039.

The standard deviation is useful for describing a set of measurements when the data has a normal or mound shape. When the data looks like this:

we expect about 68 percent of the data to fall within a distance of one standard deviation from the mean, and we expect about 95 percent of the data to fall within two standard deviations of the mean.

To illustrate this rule, recall that the OBP numbers for the AL regulars had an approximate normal shape. Also we computed the mean to be .356 and the standard deviation is .039. So we expect about

68% of the OBP numbers to fall between (.356 − .039) and (.356 + .039)

and

95% of the OBP numbers to fall between [.356 − 2(.039)] and [.356 + 2(.039)]

---

[1] To compute the standard deviation, we first find the difference of each data value from the mean, and then square each difference. Then the standard deviation is calculated by computing the sum of squared differences, dividing the sum by {the number of observations minus 1}, and then taking the square root of the result.

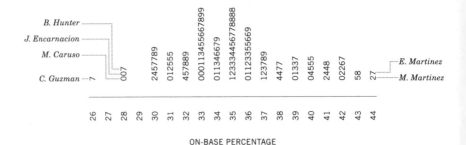

FIGURE 2-8   Stemplot of OBPs with some interesting values identified.

We compute these two intervals to be (.317, .395) and (.278, .434). Checking the data, we see that 71 out of 108 (66 percent) of the OBPs fall between .317 and .395, and 103 out of 108 (95 percent) of the OBPs fall between .278 and .434.

Interesting Values

We observe the general shape of the OBPs, looking for a typical OBP and considering the spread of the values, and then look for *interesting OBP values* that stand apart from the large cluster of OBPs in the middle. Obviously, we're interested in the largest OBPs—in 1999, Edgar Martinez had a .447 on-base percentage, followed closely by Manny Ramirez at .442. But we're also interested in unusually small values. In the redrawn stemplot of Figure 2-8, we see four small OBPs are separated from the remainder of the data. We might wonder why Guzman is an AL regular when he is only getting on base about 26 percent of the time. Perhaps these players with low OBPs are regulars on the basis of their defensive ability rather than their hitting ability.

Comparing Groups

Suppose we are interested in comparing two batches of OBPs. To do this, we first describe a way of summarizing a single batch of data using a few key numbers, and then describe a graph of these summary numbers, called a *boxplot,* which is useful in comparing groups of data.

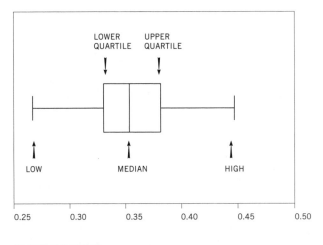

ON-BASE PERCENTAGE

FIGURE 2-9   Boxplot of OBPs for 1999 American League regular players.

A Five-Number Summary

To summarize the batch of OBPs, we can use five numbers—the median, the lower and upper quartiles, and the low and high numbers. For obvious reasons, we call these values a *five-number summary*.

Low = .267, Lower Quartile = .330, Median = .354,
Upper Quartile = .383, High = .447

A Boxplot

A boxplot is a graph of a five-number summary. To construct a boxplot, we draw a box, where the locations of the sides of the box correspond to the quartiles, and put a line in the middle corresponding to the median. We then draw lines out from the box to the low and high values. The boxplot of the OBPs in Figure 2-9 shows that a majority of the values fall in the mid 300s, with a range of about .250 to about .450.

Boxplots to Compare Groups

Boxplots are typically used to compare different groups of data. To illustrate, suppose we're interested in comparing the OBPs for the American League regular players against the OBPs for the National League regulars. There were 105 National League players who had at least 400 plate appearances (NL pitchers

Low	Lower Quartile	Median	Upper Quartile	High
0.292	0.336	0.362	0.383	0.458

TABLE 2-4   Five-Number Summary of the OBPs of the NL Regulars

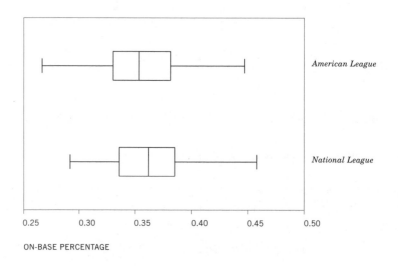

ON-BASE PERCENTAGE

FIGURE 2-10   Parallel boxplots of OBPs for 1999 regular players from the American and National Leagues.

do bat, but are naturally excluded from this list because they don't play in enough games). Table 2-4 gives the five-number summary of the 105 regular NL players' OBPs.

In Figure 2-10, boxplots of the American League and National League OBPs are drawn on the same scale. The distributions of OBPs for the two leagues look remarkably similar—the AL and NL boxes have approximately the same center and spread. Looking carefully at the two boxplots, we see that the NL OBPs are a little higher, on average, than the AL OBPs. Looking at the medians, we see that the NL median OBP was .362, compared to a median OBP of .354 for the AL. So it appears that the NL players were generally a little more successful than the AL players in getting on base in 1999.

	Low	Lower Quartile	Median	Upper Quartile	High
Offensive positions	0.280	0.340	0.358	0.386	0.447
Defensive positions	0.267	0.328	0.340	0.369	0.438

TABLE 2-5   Five-Number Summaries of the OBPs for the 1999 AL players in Offensive and Defensive Fielding Positions

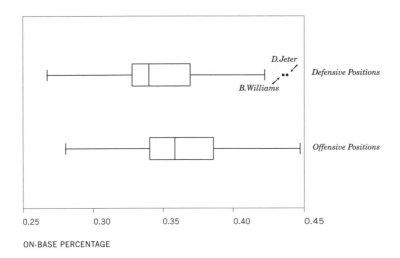

FIGURE 2-11   Parallel boxplots of the OBPs of the 1999 AL players in offensive and defensive fielding positions.

OBPs of Offensive and Defensive Players

Remember our earlier comment about the few low American League OBP values in our dataset? We speculated that these players were in the regular lineup due to their fielding rather than hitting ability.

Let's check this out with our AL data. Of the 108 regular players, 60 played in the less-important fielding positions (1B, 3B, LF, RF), and 48 played in the more important fielding positions (C, 2B, SS, CF). (Below we will call these the "offensive" and "defensive" positions, respectively.) Table 2-5 gives the five-number summary for the OBPs for each of the two groups of players, and Figure 2-11 shows parallel boxplots of the OBPs for the two groups.

Several interesting observations can be made from Table 2-5 and Figure 2-11. First, the offensive-position players tended to have OBPs 10 to 20 points higher than the defensive-position players. Specifically, the median of the offensive position players' OBP (.358) was 18 points higher than the median of the defensive position players' (.340). This substantiates the belief that many players are in the lineup for their fielding ability, not their hitting. There are, however, exceptions to this pattern. Note that there are two bullets (•) to the right of the boxplot for defensive players. Using a standard rule for determining remarkable values,[2] these OBPs were determined to be unusually high. Most fans would consider Jeter and Williams extraordinary—both play at defensive positions and are very effective hitters.

Relationships Between Batting Measures

We have spent a lot of time talking about a single measure of hitting performance—that is, a player's ability to get on base. But there is a second dimension

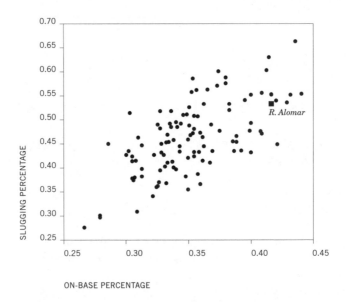

ON-BASE PERCENTAGE

FIGURE 2-12   Scatterplot of OBP and SLG for 1999 American League regulars.

[2] To determine these unusual values, one first computes a STEP which is equal to 1.5 times the distance between the upper and lower quartiles. Then one computes a LOWER FENCE which is equal to the lower quartile minus a STEP, and a UPPER FENCE which is equal to the upper quartile plus a STEP. Any data items that are smaller than the LOWER FENCE or are greater than the UPPER FENCE are called outliers which may deserve special attention.

to hitting, namely a batter's ability to advance runners already on base. The classical measure of a hitter's advancement ability is the slugging percentage, which is computed by dividing the total number of bases of all base hits by the number of at-bats. If 1B, 2B, 3B, and HR stand respectively for the number of singles, doubles, triples, and home runs of a hitter, then the slugging percentage, abbreviated SLG, is computed as follows:

$$SLG = \frac{(1 \times 1B) + (2 \times 2B) + (3 \times 3B) + (4 \times HR)}{AB}$$

Relating OBP and SLG

How is a player's on-base percentage related to his slugging percentage? A basic graph to explore the association between two variables is a *scatterplot*. For each player, we have two measures—his OBP and his SLG. For example, in 1999 Roberto Alomar had an OBP of .422 and an SLG of .530.

In Figure 2-12, we plot the ordered pair (.422, .533), which is represented by a solid dot. If we plot the ordered pair (OBP, SLG) for the 107 other regular AL players, we get the remaining points in the graph.

Looking at Figure 2-12, we see a general increasing pattern—the points drift up as one moves from the left-hand side of the graph to the right-hand side. The general conclusion from looking at this graph is that players who have high on-base percentages tend to have high slugging percentages, and players who don't get on base frequently also have low slugging percentages. This conclusion makes sense, since base hits have a positive effect on both a batter's OBP and his SLG.

Relating OBP and Isolated Power

Since a batter's OBP and SLG seem pretty highly correlated, it would seem desirable to develop an alternative measure of a hitter's ability to advance runners that is not confounded or confused with his ability to get on base. There is a measure, called *isolated power (IP)*, that is designed to do exactly that. One computes IP by subtracting a batter's batting average (AVG) from his slugging percentage (SLG).

$$IP = SLG - AVG = \frac{2B + (2 \times 3B) + (3 \times HR)}{AB}$$

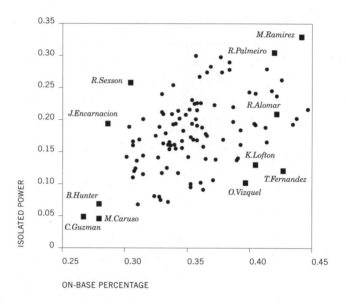

FIGURE 2-13   Scatterplot of OBP and IP for 1999 American League regulars.

Figure 2-13 shows a scatterplot of the isolated power values against the on-base percentages for the American League regulars. Note that we see the same type of pattern in this graph as we saw in our first scatterplot, indicating that players who have high OBP values tend also to have high IP values (and players who have low OBP values also have low IP values). But the relationship between OBP and IP appears weaker than the relationship between OBP and SLG, indicating that we have partially succeeded in developing two hitting statistics that aren't strongly linked. In other words, IP appears to measure a player's ability to get extra bases and advance runners that is distinct from his ability to get on base.

This scatterplot of OBP and IP is useful in describing different types of baseball hitters. The players in the lower right part of the graph, such as Kenny Lofton, Tony Fernandez, and Omar Vizquel, are hitters who are successful in getting on base, but have relatively little power to get extra base hits. In contrast, the hitters in the upper left section of the plot, such as Richie Sexson and Juan Encarnacion, have good power (indicated by high IP values), but relatively poor ability to get on base. Obviously the most valuable hitters are the ones who have high values of both OBP and IP. Manny Ramirez, the most extreme point in the upper-right section of the plot, had great values of both OBP and IP in 1999, and finished tied for third in the AL MVP voting.

Statistic	Description	Statistic	Description
P	pitching arm	HR	home runs allowed
GP	games pitched	TB	total bases allowed
GS	games started	ER	earned runs
W	wins	ERA	earned run average
L	losses	IP	innings pitched
SV	saves	SO	strikeouts
CG	complete games	BB	base on balls
S	shutouts	BK	balks
R	runs	HP	batters hit by pitch
H	hits		

TABLE 2-6   1999 Statistics for Major League Pitchers

What about Pitching Data?

We've learned some basic techniques for graphing and summarizing hitting data. The same techniques can be used to analyze any batch of baseball data, including the statistics used to evaluate pitchers. The Major League Baseball website provides a number of statistics for pitchers; Table 2-6 describes these pitching statistics.

We learn quite a bit about a pitcher by exploring these statistics. The games pitched (GP), games started (GS), and innings pitched (IP) tell us how active the pitcher was during the season. The wins (W), losses (L), and saves (SV), are direct measures of the success and failure of a pitcher, since the objective of a team is to win games. Indirect measures of success are statistics such as runs (R), hits (H), and home runs allowed (HR), since the hits, runs, and home runs allowed by a pitcher are positively correlated with an *opponent's* success. Pitchers are usually compared using their win/loss records and their earned run averages (ERAs). An ERA is the average number of runs allowed by a pitcher (not counting runs due to miscues by his teammates) for a nine-inning game. An interesting question is whether an ERA is the best way to evaluate pitching performance. (A general discussion on rating players is covered in Chapter 6.)

Strikeouts and Walks

Here we'll explore two basic pitching statistics, the number of strikeouts and the number of walks (bases-on-balls) for the 1999 National League pitchers. Strikeouts and walks are interesting events in baseball. When a pitcher gets a strikeout, one gets the impression he is dominant, dictating from the mound. However, a strikeout produces only a single out, and it is not clear that a pitcher who throws a lot of strikeouts will be effective in not allowing runs and ultimately winning games. When asked to name the ultimate strikeout pitcher, a lot of people would think of Nolan Ryan, but his lifetime record of wins was only about 53 percent. (Of course, one could argue that that this was due at least in part to the poor teams on which he played.) Similarly, when a pitcher walks a batter, one thinks that he has lost his control and given up an easy on-base. So it doesn't seem desirable to walk many batters, but it is not entirely clear what impact a walk has on the opposing team scoring runs. In the remainder of this chapter, we'll explore strikeout and walk statistics to address the following:

- What is a typical strikeout rate (or walk rate) among pitchers? That is, how many strikeouts (or walks) does a pitcher typically get for nine innings?

- Are there unusually good or unusually poor pitchers relative to striking out hitters? Likewise, are there pitchers with unusually good control who don't walk many batters?

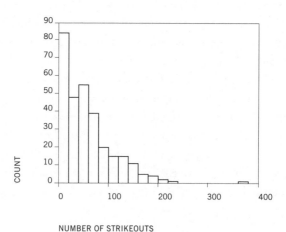

FIGURE 2-14   Histogram of number of strikeouts of 1999 National League pitchers.

- Do starting pitchers strike out more hitters than relief pitchers? How do starting and relief pitchers differ with respect to walking hitters?

Looking at Strikeout Totals

The MLB website gives 1999 pitching statistics for 300 NL pitchers. Figure 2-14 displays a histogram of the season strikeout totals for these 300 pitchers. This histogram has a distinctive shape:

which we call *right-skewed*. Most of the strikeout totals fall close to zero, and there are relatively few large strikeout totals. This histogram shape is very common when our statistic is a *count* of something. If we graph the counts of hitters' home runs, the counts of walks given up by pitchers, the counts of errors made by third basemen, or the counts of games won by pitchers, we will find that the shape of the data will be right-skewed. Most of the data will be clustered at small values, and there will be a few large numbers.

Why does the histogram in Figure 2-14 have this right-skewed shape? Of the 300 pitchers in this list, many have pitched few innings and have recorded only a small number of strikeouts. Figure 2-15 shows a histogram of the total innings pitched. We see three humps in the histogram—there are many pitchers in this

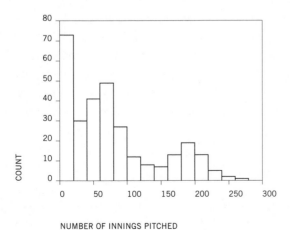

NUMBER OF INNINGS PITCHED

FIGURE 2-15 Histogram of number of innings pitched by 1999 National League pitchers.

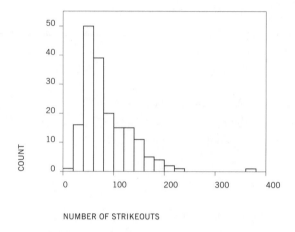

FIGURE 2-16   Histogram of number of strikeouts of 1999 National League pitchers with at least 50 innings pitched (180 pitchers).

group who pitched from 0–20 innings, there is another large clump of pitchers (primarily relievers) who pitched from 50–70 innings, and a clump of pitchers (starters) who pitch around 200 innings. So there are many part-time pitchers with few innings pitched, and the statistics for these part-timers are clouding the picture of the strikeout data. It seems better to look only at pitchers who have appeared in a minimum number of innings—and we'll arbitrarily set this minimum at 50.

If we remove the pitchers with fewer than 50 innings pitched, we're left with 180 pitchers. A histogram of the season strikeout totals for these NL pitchers is shown in Figure 2-16. We see a better picture of the strikeout totals—the shape of the data is still right-skewed, with a large number of pitchers having from 40–100 strikeouts, and a few pitchers with a large number of strikeouts. But the data from pitchers who appeared in only a very few innings—perhaps because they were injured, or sent back down to the minors—no longer has a significant effect on the graph.

Defining a Strikeout Rate

The strikeout king in baseball is traditionally viewed as the pitcher with the greatest number of strikeouts. That this is so should come as no surprise, but it should also be pretty obvious that having the greatest number of strikeouts in a season is not the best measure of a pitcher's ability to strike out hitters. All you

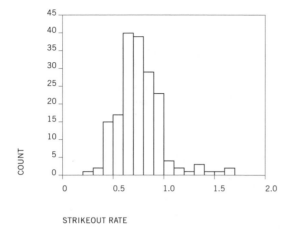

FIGURE 2-17   Histogram of strikeout rates (strikeouts per inning pitched, or SOR) of 1999 National League pitchers with at least 50 innings pitched (180 pitchers).

have to do is think of those who pitch in relief, for example. In the group of pitchers who have pitched at least 50 innings, a pitcher who has taken the mound in more innings is more likely to strike out more hitters. It seems better, then, to take the average number of strikeouts per inning, which is the basis for a statistic called the strikeout-rate (SOR):

$$SOR = \frac{number\ of\ strikeouts}{innings\ pitched} = \frac{SO}{IP}$$

Figure 2-17 shows a histogram of the SOR values for our 180 NL pitchers. The shape of this histogram, which is normal, or bell-shaped, is typical for a *derived baseball statistic*—this is, a statistic that is derived as a ratio of basic counts. We saw this same data shape in our exploration of OBPs. We would expect to see a similar normal shape for other derived statistics—such as ERA, BVG, or SLG—computed for players who have appeared in a reasonable number of innings.

Here we see that a typical strikeout rate for a 1999 NL pitcher is about .7 per inning, or about $9 \times .7 = 6.3$ strikeouts for every 9 innings pitched. We next look for unusual strikeout rates, statistics that seem markedly different from the average. In Figure 2-17, we notice 15 pitchers with strikeout rates exceeding 1, and three pitchers exceeding a rate of 1.5. This is interesting, and deserves a closer look. In Figure 2-18, we show these strikeout rates as a stemplot, identifying the pitchers with the highest values.

2	7
3	77
4	00124455578899
5	000012233333478899
6	0000000111222344444455556677778888899999
7	00001111122233334445566667778888889999
8	000001111333345555566677888899
9	0000111222345566667799
10	02678
11	47
12	3
13	024 *M.Williams, U.Ubrina, R.Johnson*
14	4 *J.Rocker*
15	2 *M.Mantei*
16	47 *A. Benetez, B.Wagner*

FIGURE 2-18   Stemplot of strikeout ratios of 1999 NL pitchers, including names of pitchers with unusually large values.

STARTERS		*RELIEVERS*
	2	7
7	3	7
988755544210	4	09
9887433333210	5	00029
9999988877765554444422110000	6	00123456788
99988766544333221110000	7	1123456677888899
998755433100	8	00011133555667888
975433110	9	00012225666679
	10	02678
	11	47
	12	3
R. Johnson 4	13	02 *M. Williams, U. Ubrina*
	14	4 *J. Rocker*
	15	2 *M. Mantei*
	16	47 *A. Benetez, B. Wagner*

FIGURE 2-19   Back-to-back stemplots of strikeout rates of 1999 NL starters and relievers.

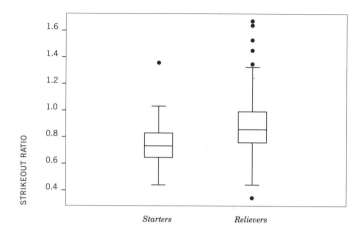

FIGURE 2-20   Parallel boxplots of strikeout rates of 1999 NL starters and relievers.

Comparing Strikeout Rates of Starters and Relievers

Of the seven pitchers with the highest strikeout rates, we note that only one (Randy Johnson) is a starter. That raises the question—do relievers typically strike out more batters than starters? To answer this question, we divide the 180 NL pitchers into two groups, defining a "starter" as a pitcher who has started at least ten games in the 1999 season.

A useful graph for comparing the strikeout rates of starters and relievers is the *back-to-back stemplot*, shown in Figure 2-19. We put the stems in the middle; the leaves for the starters go to the left, the leaves for the relievers to the right. As before, we identify only those pitchers with high strikeout rates.

Another way to compare these two groups of strikeout rates is by use of boxplots (see Figure 2-20).

We see some interesting things from the displays of stemplots and boxplots.

1. Generally, the NL relievers seem to strike out more batters per inning than the NL starters. The median strikeout rate of the starters was .68, compared to a rate of .85 for the relievers. In nine innings, a typical reliever will strike out 1 more batter (7.3) than a typical starter (6.1).

2. The spread in strikeout rates among the relievers is greater than the spread in rates for the starters. The lower and upper quartiles

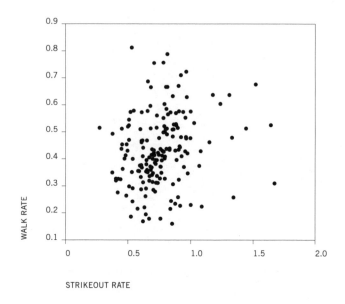

FIGURE 2-21   Scatterplot of strikeout and walk rates of 1999 NL pitchers.

for the relievers is .70 and .96, compared to quartiles of .59 and
.79 for the starters.

3. The boxplot display identifies several pitchers with unusually low
   or high strikeout rates. By breaking the pitchers into the two
   groups, only one pitcher seems to have a strikeout rate that
   clearly stands out from the rest. With apologies to Curt Schilling,
   there was no starting pitcher in the NL in 1999 who had a
   strikeout rate even close to Randy Johnson's.

Association Between Strikeouts and Walks?

As we move our discussion from strikeouts to walks, it is reasonable to ask if
there is any relationship between the two statistics. Nolan Ryan is widely (and
correctly) known as a pitcher who struck out many batters but also gave up a lot
of walks. (Looking at his career statistics, we notice that Ryan had 5714 strikeouts and 2795 walks in 5386 innings; his strikeout rate was 1.07 and his walk
rate was .52, both of which appear to be large.) Are these large strikeout and
walk rates typical of a fastball pitcher? If so, one might expect a positive associ-

```
1 | 667789
2 | 112222344
2 | 55667788999
3 | 0111111222222233344444
3 | 5555555666777777788899999
4 | 00000011111112222223333333333444
4 | 55666667777788899999
5 | 00111111111222222234
5 | 556667777777789
6 | 002333
6 | 666778
7 | 12
7 | 558
8 | 1
```

FIGURE 2-22   Stemplot of walk rates of 1999 NL pitchers.

ation between walks and strikeouts. To check this out, we first compute the walk rate for each AL pitcher, as shown in the following equation, and then construct a scatterplot of SR against WR, as shown in Figure 2-21.

$$WR = \frac{number\ of\ walks}{innings\ pitched} = \frac{BB}{IP}$$

There does not seem to be much of a pattern in the graph. There is a slight drift of the point cloud from lower left to upper right, but at best, there is a weak positive association between strikeout and walk rates.

Exploring Walk Rates

How often do pitchers walk batters? Figure 2-22 shows a stemplot of the walk rates for the NL pitchers with at least 50 innings pitched. (We've ignored the decimal point in presenting this stemplot.)

We see basically the same data shape as we saw earlier for strikeouts. The shape of the walk rates is slightly right-skewed—the median value is .42, with lower and upper quartiles of .35 and .51. So half of all the pitchers walk between .35 and .51 batters per inning, which translates to between 3 and 4.5 walks per game. There is a sizeable range in the data from .16 (Shane Reynolds) to .81 (Steve Avery).

	STARTERS			RELIEVERS
S. Woodard, J. Lima, G. Maddux, S. Reynolds	9766	1	78
	443211	2	222
	9998876655	2	7
	4444433322221111	3	011222
	98887776665555555	3	777779999
	43333333322211100000	4	0111122233344
	977665	4	56667778889999
	1111	5	0011111222222234
	777776655	5	677789
	30	6	0233
	86	6	6677
		7	12
J. Sanchez, J. Bere	85	7	5 W. Gomes
S. Avery	1	8	

FIGURE 2-23   Back-to-back stemplots of walk rates of 1999 NL starters and relievers.

Comparing Walk Rates of Starters and Relievers

We conclude this chapter by addressing the question: Do starting pitchers and relievers have different tendencies to walk batters? As before, we divide the pitchers into starters (at least 10 starts in 1999) and relievers. Back-to-back stemplots are shown in Figure 2-23.

Several interesting features are noticeable. Relievers appear to walk more hitters than starters. The median number of walks per inning for relievers is .48, compared to a median of .38 for starters. (Over nine innings, the relievers generally walk about an additional batter per inning.) Although starters generally exhibit more control, there are three notable exceptions—J. Sanchez, J. Bere, and S. Avery—who appear to have unusually little control. Being Phillies fans, we're a bit distressed that Wayne Gomes, their closer in 1999, had the worst walk rate among all the NL relievers, but we don't altogether lose hope: Mitch Williams (even though he gave up the Joe Carter home run in the 1993 World Series) had a good relief pitching year for the 1993 Phillies despite walking a lot of batters.

INTRODUCING Probability

CHAPTER 3

In Chapters 1 and 2, we discussed some methods for exploring baseball statistics. In Chapter 1, we focused on the relatively simple statistical models used in some popular tabletop games. In Chapter 2, we looked at graphical and descriptive methods that, while simple, are the fundamental tools and models used by all statisticians. In particular, Chapter 2 used the example of on-base percentage (OBP) data to describe and analyze the performance of AL hitters. With this basic knowledge, we know what's being said, statistically speaking, when someone declares, "Roberto Alomar had a .422 on-base percentage in 1999." We know that the value .422 is large relative to the entire distribution of OBPs. (See the stemplot of 1999 AL OBPs in Figure 2-3 of Chapter 2.)

Now that we have explored the OBP data and made a few graphs and charts, are we done? Are professional statisticians primarily interested in numbers and patterns of data as they are revealed in summaries and graphs? Aren't statisticians, after all, just glorified number-crunchers or graph makers? Obviously, we don't think so. What the serious statistician is really interested in are the *conclusions* or *inferences* that one can draw from the data.

Beyond Data Analysis

Let's contrast the intent of this baseball book with practically any other baseball statistics book that is published each spring. Essentially baseball statistics books fall in two categories. One type, which we will call the "numbers book," just presents tables and tables of baseball data. This book is designed for the fan who wants to know all of the statistics from his or her favorite player. The goal of this type of book is just to *tabulate* numbers—to put numbers in tables—in a form that's convenient for the fan to retrieve and review.

A second type of baseball statistics book—we'll call it the "analysis book"—tries to go one step further. It will ask an interesting baseball question, then present relevant data to answer it: "Who was the better lead-off hitter in 1999: Chuck Knoblauch or Kenny Lofton?"

Since a lead-off hitter is supposed to get on base, the analysis book would focus on OBPs, obviously the relevant data set. Knoblauch's and Lofton's numbers are highlighted in Table 3-1, which shows a small segment of a straightforward alphabetical list of regular AL players with their 1999 OBP results.

The book notes that Lofton was more successful in getting on base than Knoblauch, as his OBP value was 12 points higher. Since the book doesn't say anything else, the reader is left to draw the conclusion that Lofton is better than Knoblauch at getting on base. But does Lofton *really* have more ability to get on base than Knoblauch?

This question is different from the one posed by the analysis book, which focused on performance, or results, for 1999. We know that Lofton had a better year—we are not disputing the calculation of his 1999 OBP value. But did Lofton really have a greater *ability* to get on base than Knoblauch in 1999? When we say ability, we are referring to the characteristics of a hitter, such as

Player	1999 OBP
D. Jeter	0.438
C. Johnson	0.340
D. Justice	0.413
C. Knoblauch	**0.393**
M. Lawton	0.353
D. Lewis	0.311
K. Lofton	**0.405**
T. Martinez	0.341
D. Martinez	0.461
E. Martinez	0.447
F. McGriff	0.405

TABLE 3-1   1999 OBPs for a Selection of AL Players

his batting stroke, his eye for watching pitches carefully, and his patience, all of which would contribute to his ability to reach base.

There are two possible explanations (at least) for the difference between Knoblauch's .393 and Lofton's .405 . Maybe Lofton really *is* better at getting on base than Knoblauch, and the 12-point difference is just a reflection of this fact. But maybe the two players have the same abilities to get on base, or perhaps Knoblauch has a superior ability, but by luck or chance Lofton got a better OBP value in 1999. Which explanation is right? The professional statistician's job is to help distinguish differences due to *real effects* from differences that can be explained by *chance*.

Looking for Real Effects

In the course of writing this book, we had the opportunity to pose the Knoblauch-Lofton question to a nine-year-old baseball fan, showing him the OBP data in Table 3-2. We pointed out that although Lofton had a higher OBP in 1999, the difference could be due to luck—maybe Lofton was more lucky than Knoblauch this season. But the boy, looking at the table, said that Lofton must be better—especially since he also had a higher OBP in 1998.

He had a point, and a good point. If one player has a better on-base ability than a second player, then one would expect the first player to get generally a

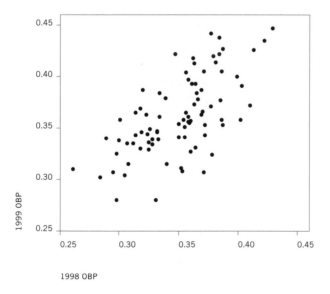

FIGURE 3-1   Scatterplot of 1998 and 1999 OBPs for AL regular players.

higher OBP value across seasons. The question is: how dependable is a high OBP in one year as a predictor for a high OBP in the next?

Let's explore the relationship between a player's OBP in 1998 and his OBP in 1999. Table 3-2 shows the OBPs for the players who had at least 400 plate appearances in 1999 and had at least 400 appearances in the previous year; Figure 3-1 shows a scatterplot of the same data.

Player	1999 OBP	1998 OBP	Player	1999 OBP	1998 OBP	Player	1999 OBP	1998 OBP
R. Alomar	0.422	0.347	D. Fletcher	0.339	0.328	P. O'Neill	0.353	0.372
G. Anderson	0.336	0.325	N. Garciaparra	0.418	0.362	J. Offerman	0.391	0.403
B. Anderson	0.404	0.356	J. Giambi	0.422	0.384	M. Ordonez	0.349	0.326
B. Ausmus	0.365	0.356	J. Gonzalez	0.378	0.366	R. Palmeiro	0.420	0.379
D. Bell	0.331	0.364	T. Goodwin	0.324	0.378	D. Palmer	0.339	0.333
A. Belle	0.400	0.399	S. Green	0.384	0.334	J. Posada	0.341	0.350
M. Bordick	0.334	0.328	R. Greer	0.405	0.386	M. Ramirez	0.442	0.377
S. Brosius	0.307	0.371	B. Grieve	0.358	0.386	J. Randa	0.363	0.323
M. Cairo	0.335	0.307	K. Griffey	0.384	0.365	A. Rodriguez	0.357	0.360
J. Canseco	0.369	0.318	B. Higginson	0.351	0.355	I. Rodriguez	0.356	0.358
M. Caruso	0.280	0.331	B. Hunter	0.280	0.298	T. Salmon	0.372	0.410
T. Clark	0.361	0.358	B. Huskey	0.338	0.300	D. Segui	0.355	0.359
R. Clayton	0.346	0.319	D. Jeter	0.438	0.384	M. Stairs	0.366	0.370
R. Coomer	0.307	0.295	C. Johnson	0.340	0.289	M. Stanley	0.393	0.364
M. Cordova	0.365	0.314	D. Justice	0.413	0.363	S. Stewart	0.371	0.377
D. Cruz	0.302	0.284	C. Knoblauch	0.393	0.361	B. Surhoff	0.347	0.332
J. Cruz	0.358	0.354	M. Lawton	0.353	0.387	M. Tejada	0.325	0.298
J. Damon	0.379	0.339	D. Lewis	0.311	0.352	F. Thomas	0.414	0.381
R. Davis	0.304	0.305	K. Lofton	0.405	0.371	J. Thome	0.426	0.413
C. Delgado	0.377	0.385	T. Martinez	0.341	0.355	J. Valentin	0.315	0.34
R. Durham	0.373	0.363	E. Martinez	0.447	0.429	M. Vaughn	0.358	0.402
D. Easley	0.346	0.332	F. McGriff	0.405	0.371	O. Vizquel	0.397	0.358
D. Erstad	0.308	0.353	M. McLemore	0.363	0.369	T. Walker	0.343	0.372
T. Fernandez	0.427	0.387	B. McRae	0.327	0.360	B. Williams	0.435	0.422
J. Flaherty	0.310	0.261	T. O'Leary	0.343	0.314	T. Zeile	0.354	0.350

TABLE 3-2   Two-Year On-Base Percentages for 1999 AL Regular Players

	1999 OBP	*1998 OBP*
M. Bordick	0.334	0.328
S. Brosius	0.307	0.371

TABLE 3-3   1998 and 1999 OBPs for Mike Bordick and Scott Brosius

Looking at the scatterplot in Figure 3-1, we see a positive relationship. This means that players who got high OBP values in 1998 tended also to get high OBP values in 1999. (A positive relationship also means that players who got low OBP values in 1998 tended to get low OBP values in 1999.) The scatterplot and the data from Table 3-2 provide evidence of year-to-year OBP consistency. This implies that underlying the OBPs observed in each year for a player resides a consistent OBP ability. If players all had the same ability to get on base, and differences between season OBPs between players were only due to luck, then there wouldn't be any pattern in the scatterplot in Figure 3-2.

That said, though we do see a relationship in the scatterplot between 1998 and 1999 numbers, it is *not* a strong relationship. In fact, you can find pairs of players, such as Mike Bordick and Scott Brosius (see Table 3-3) where one player (Brosius) had a higher OBP in 1998 and the other player (Bordick) had a higher OBP in 1999.

Predicting OBPs

While we're on the subject of comparing OBPs for two different years, let's pose another question. Suppose that you know a player's OBP in 1998. What is the best prediction of his OBP the next year?

This question can be answered by revisiting our scatterplot of the 1998 and 1999 OBP values in Figure 3-1. When we see a relationship in a scatterplot, it is helpful to summarize this relationship by means of a straight line that passes through the points. The equation of this straight line gives a simple formula that relates a player's 1999 OBP value with his 1998 value. In statistics, there is a basic recipe (called *least squares*) which finds the equation of the line which is the "best fit" through the points of a scatterplot. In this case, the best line has the following formula:

$$OBP_{1999} = .0946 + .761 \, OBP_{1998}$$

This line is drawn on the scatterplot in Figure 3-2. This best line can be used to predict a player's 1999 OBP if you know his 1998 OBP, and these predictions are a bit surprising. Let's illustrate.

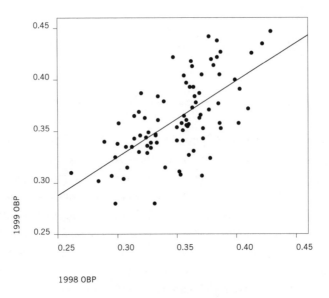

FIGURE 3-2   Scatterplot of 1998 and 1999 OBPs for AL regular players with a "best line" drawn on top.

Kenny Lofton had a .371 OBP in 1998, when the average OBP value (among AL regulars) was .350, so Lofton was *21 points better than average* in 1998. It would be reasonable to predict that Lofton would also be 21 points better than average in 1999. However, the prediction using the best line formula is:

$$OBP_{1999} = .0946 + .761 \, OBP_{1998}$$

and, in Lofton's case, the best line prediction is:

$$OBP_{1999} = .0946 + .761 \, (.371) = .377$$

which is only *15 points better* than the 1999 average value of .362.

Likewise, if you use the best-line formula to predict any other player's 1999 OBP, you'll discover that the 1999 prediction is closer to the 1999 average than the 1998 OBP is to the 1998 average. What is going on?

This illustrates a general result, called "regression to the mean," which applies to any baseball statistic that is measured for two years in a row. (Actually it applies to many situations besides baseball.) In this setting, it means that a hitter's OBP tends to be closer to the average in the second year than it does for the first year. The phenomenon is also called the "sophomore slump," and certainly there is no shortage of media attention for a ballplayer who has a relatively mediocre second season after a spectacular rookie season.

But this relatively weak second year illustrates this regression to the mean effect—ballplayers who have extremely higher-than-average seasons one year tend to have less extreme seasons the next.

Probability Models

By comparing the OBPs of AL regulars for two consecutive years, we now believe that players indeed have different abilities to get on base. But how can we describe a player's ability to get on base? Fans use expressions like, "He's good at getting on base," or "He has good bat control," to describe a player's hitting ability. Can we use numbers instead of words to explain a hitter's ability?

Statisticians use numbers assigned to chance outcomes, *probabilities*, to draw conclusions from data. We know that many things in life are uncertain. We don't know in advance the outcome of a coin toss, the value of the Dow Jones Industrial Average at the end of next month, the winner of the 2010 World Series, or for that matter the year in which the Phillies will win their next World Series. But the statistician recognizes that, although many aspects of life are uncertain, there exist general patterns amid this uncertainty, and probability is a method for describing those general patterns.

To understand a hitter's ability to get on base, the statistician constructs a *probability model,* or a *model* for short. A model is a description of a random process that could possibly generate the baseball data. (We have already described several models in Chapter 1 that underlie tabletop baseball games.) Let's look at a simple example. In 1999, Roberto Alomar had 682 plate appearances and got on base 288 times, for an OBP of 288/682 = .422. To investigate how a model works, we will think of a simple random experiment that could have produced Alomar's data. Before we do that, though, we'll take a quick side trip to look at a model we're all familiar with, to get a sense of the variability in "chance" outcomes.

A Coin-Toss Model

Consider the simple experiment of tossing a coin. We are thinking of the usual two-sided coin, where the chance of throwing heads is the same as the chance of throwing tails. In coin-tossing, the chance of tossing heads is .5—we know this since we believe that it's equally likely to land heads or tails. The number .5 represents the *true proportion* of heads—this is the fraction of heads that we expect to get if we toss the coin repeatedly. We can think of .5 as an attribute of the coin, which comes from our belief that the coin is fair. To put it another way, we can say that the coin's ability to land heads is .5. Our coin-tossing model simulates a perfect 50-50 split between heads and tails.

What happens, though, when we turn from the model, which is a mathematical formulation, and actually start tossing a coin? Let's contrast the proportion of heads described by the model with the proportion of heads observed in an actual series of tosses. Suppose we toss this fair coin 10 times and we observe 7 heads and 3 tails. The observed proportion of heads is equal to:

$$\frac{7}{10} = .7, \text{ or } 70\%$$

Since we got 70 percent heads, does that mean that the coin is not fair? No. Or does it mean that our model is invalid? No, again. The coin is fair, and the model is valid, but we were lucky (or unlucky) to get 7 heads in this particular set of actual tosses. In fact, it is entirely reasonable to get 7 heads out of 10 tosses. What we have to do in this situation is distinguish between the observed proportion in our data, which is 70 percent, and the so-called *true proportion*, which remains 50 percent if the coin is fair.

Is it possible for the observed proportion of heads in 10 tosses to be equal to the true proportion of 50 percent? Yes, certainly, one could get 5 heads in 10 tosses. But this won't typically be the case when you have 10 tosses. In fact, it is *more likely* in 10 tosses that the observed proportion will be *different from* the true proportion.

Let's consider an illustration of this idea. On a computer, we simulated the experiment of tossing a fair coin 10 times. Repeating the experiment 100 times, and keeping track of the number of observed heads in each of these 100 cycles, we arrived at the outcomes summarized in Table 3-4. In Table 3.5, we used the same data to record the distribution of outcomes according to whether a 10-toss series resulted in 1 heads, 2 heads, 3 heads, and so on, along with a calculation of the probability for each of these outcomes.

We see from Table 3-5 that we observed a 5-heads outcome in 30 of 100, or 30 percent, of the experiments. So the chance that we would observe exactly 50 percent heads in our 10 tosses is only about .3. Saying this a different way, the chance that the observed proportion is different from the true proportion is .7. So while our coin-toss model is valid, there is considerable variation in the results of our 10 tosses of our coin.

There is one situation where the observed proportion will be very close to the true proportion. Suppose that we were able to toss the coin a very large number of times—say, a million. In such a case, the observed proportion of heads will be

6	4	6	3	5	5	7	6	5	7
5	7	5	4	5	6	5	4	6	5
3	4	3	7	4	6	8	5	3	5
6	5	4	4	5	7	2	5	5	4
6	4	5	4	5	8	3	4	1	5
6	4	5	5	4	3	5	4	6	3
4	5	6	7	6	6	3	7	3	5
7	3	4	6	4	5	6	6	7	5
4	2	5	5	4	3	5	5	5	6
4	2	6	7	7	5	8	4	4	4

TABLE 3-4   Number of Heads in 10 Tosses for 100 Experiments

Number of heads	1	2	3	4	5	6	7	8
Count	1	3	11	23	30	18	11	3
Proportion	0.01	0.03	0.11	0.23	0.3	0.18	0.11	0.03

TABLE 3-5   Probabilities for Number of Heads in 10 Tosses of a Fair Coin

very close to the true proportion of 50 percent. In fact, we can state a general rule: as we toss the coin more and more times, the observed proportion will generally get closer and closer to 50 percent.

Observed and True OBPs

Let's turn our discussion from coins back to baseball, and look again at the batting behavior of Roberto Alomar in 1999. Recall that Alomar came to bat 682 times and got on base 288 times:

$$\frac{288}{682} = .422$$

This number is his observed OBP based on 682 opportunities to bat.

Now Alomar possesses a certain ability to get on base. This ability is based on his ability to see pitched balls, his hitting stroke, and his speed to run to first base. We can signify Alomar's ability to get on base this season by a number p, which we will call his *true on-base percentage*, or *true OBP*.

Like the coin-toss probability of 50 percent, this number represents Alomar's chance of getting on base in a single at-bat. Is the true OBP equal to .422 (his observed average in 1999)? Probably not, for the same reason that the proportion of heads in 10 tosses is unlikely to be equal to the coin's true probability.

When we toss a coin, we know the true proportion is one half. Why? Well, we know something about the composition of the coin (it has two sides, heads and tails), and we've likely had some experience tossing coins, so we believe the coin to be fair. Is it possible to know Alomar's true OBP or his ability to get on base? Not really. We will learn about Alomar's ability to get on base from observing his hitting performance for many seasons. But we don't know his true probability of getting on base for the 1999 season. If Alomar had millions instead of hundreds of plate appearances during 1999, we could come up with a very close approximation of p, his true OBP, but of course this will not happen, so p, at least for now, will remain a mystery. We can guess that his true OBP is in the .422 ballpark (pardon the pun), but it could conceivably be .380 or .440. We don't know and never will know the exact value of p.

Let's illustrate the difference between Alomar's hitting ability and his season performance by means of a simple simulation. Suppose Alomar is a true .380 OBP player—the chance that he gets on base in a plate appearance is 38 percent. We will simulate Alomar's hitting results for a season of 682 plate appearances. Here is how the simulation might work: Imagine a spinner, pictured in Figure 3-3, where the pointer of the spinner can land anywhere on the circle. If the spinner is spun and the pointer falls in the On-Base region (the emphasized area), we record an on-base event; otherwise, we record that he didn't get on base. (The size of the On-Base region, in this case, is 38 percent of the total area.) If we spin the spinner 682 times, we simulate a whole season of hitting, and the total number of pointers that fall in the On-Base region will be Alomar's number of times on base for the season.

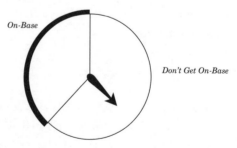

FIGURE 3-3   Spinner for simulating hitting for Roberto Alomar.

276	260	268	283	262	256	263	254	237	241
249	251	255	248	269	229	253	249	265	260
260	259	269	288	267	273	266	277	249	279
263	263	267	264	263	268	271	246	254	242
248	270	258	230	263	258	243	272	274	280
251	265	278	264	259	255	262	246	260	266
279	242	245	257	258	260	280	262	248	289
264	273	264	249	253	247	263	253	276	263
248	241	271	249	251	243	251	247	260	262
257	244	251	274	246	264	244	237	280	256

TABLE 3-6   Number of Times on Base for 100 Seasons Assuming Roberto Alomar Is a True .380-OBP Player

.405	.381	.393	.415	.384	.375	.386	.372	.348	.353
.365	.368	.374	.364	.394	.336	.371	.365	.389	.381
.381	.370	.394	.422	.391	.400	.390	.406	.365	.409
.386	.386	.391	.387	.386	.393	.397	.361	.372	.355
.364	.396	.378	.337	.386	.378	.356	.399	.402	.411
.368	.389	.408	.387	.380	.374	.384	.361	.381	.390
.409	.355	.359	.377	.378	.381	.411	.384	.364	.424
.387	.400	.387	.365	.371	.362	.386	.371	.405	.386
.364	.353	.397	.365	.368	.356	.368	.362	.381	.384
.377	.358	.368	.402	.361	.387	.358	.348	.411	.375

TABLE 3-7   OBP for 100 Seasons Assuming Roberto Alomar Is a True .380-OBP Player

Actually we didn't actually use a spinner. A computer is a more convenient tool for performing this kind of simulation, particularly if you want to run the simulation a large number of times. We did the simulation 100 times, obtaining the number of times on base for each of 100 seasons! Assuming a true OBP of .380, Table 3-6 shows the number of times on base that we observed. In Table 3-7, we change these on-base numbers to OBPs by dividing each by the number of plate appearances (682). Figure 3-4 displays these 100 season OBPs using a stemplot.

Remember that we are assuming that Alomar is truly a .380 on-base player. Also, we're assuming that he remains a true .380 on-base player for all of these

33	57
34	77
35	334466779
36	0002233335555588888
37	1112233556688899
38	111111444455555557777788
39	001133445778
40	0011446799
41	0005
42	23

FIGURE 3-4   Stemplot of 100 simulated OBPs of a true .380 hitter.

100 seasons. But we see from the stemplot that there is a lot of variation in his season batting performances. In one unlucky season, he had only a .335 OBP. At the other extreme, in one season he was very fortunate and had a .423 OBP. These simulated results demonstrate just how different a player's *seasonal* OBP can be from his *true* OBP (that is, his true probability, p). Also, it is important to note the size of the variation: there is an 88-point differential between Alomar's best (.423) and worst (.335) seasons.

Learning about Batting Ability

In the previous section we assumed that we *knew* Alomar's ability to get on base, measured as the probability p, and we found plausible values for the 100 season OBPs. As mentioned earlier, we of course don't know Alomar's real on-base probability. But can we get any closer to knowing p, given that we *do* know his 1999 seasonal OBP was .422?

We will use a simple hypothetical example to illustrate how we can learn about a player's hitting ability. Suppose a particular veteran skipper—we'll call him Casey—has been managing for thirty years. Based on his experience, Casey classifies hitters into five distinct ability categories. The "weak" hitters get on base only about 20 percent of the time, the "average" hitters get on base 30 percent of the time, the "above-average" hitters have an on-base rate of 40 percent, the "excellent" hitters 50 percent, and the "superstars" 60 percent. All of these numbers represent true OBPs; if Casey rates a hitter at 30 percent, that hitter will get on base 30 percent of the time if he is given a very large number of chances at bat. (Note that we're assuming a very broad range of abilities—a player with $p = .200$ is pretty terrible and a player with $p = .600$ is better than the best on-base men in the history of the game.)

Now suppose Casey is asked to manage a new team, and he is unfamiliar with the hitting abilities of the players in his dugout. Moreover, he is told by the owner (who knows a lot about the abilities of the players on his team) that they are equally divided between the five hitting-ability categories described above. That is, one-fifth of the players are truly weak hitters, another fifth are average, one-fifth are above-average, one-fifth are excellent, and one-fifth are superstars. There are here, of course, a lot of unlikely assumptions. First, it is not believable that an experienced manager like Casey would not immediately get to know his players; he is, after all, an experienced baseball man. And it is even more surprising that the owner knows anything about his players. And perhaps most unlikely, talent is rarely if ever evenly divided between the five ability levels; there are typically more average hitters than superstars. Nonetheless, these fairly ridiculous assumptions simplify our example considerably, and will help illustrate an important point.

Since the manager has no idea which players are good or poor in hitting, he plans to insert them randomly into the batting lineup. Casey's philosophy is simple. "I have no idea who the good hitting players are, so there is no harm in playing them randomly. But I will learn which players are good after watching them perform in a week's worth of baseball games." Here is the big question. Can the manager *really* learn much about a hitter's ability to get on base if he observes the player in 20 plate appearances, which more or less would be the total appearances for a player in a week? Specifically, suppose a particular hitter, whom we'll call Mickey, gets on base 8 times in 20 PAs for a .400 observed OBP. What has the manager learned about Mickey's true OBP?

We perform a simple simulation to illustrate the process of selecting a player at random from the dugout and having the player bat for 20 PAs. (This is not real baseball, but it is a reasonable representation for what is happening in this example.) We start with a bowl containing five spinners, one with an OBP area of .200, another with a OBP area of .300, and so on. (These spinners correspond to the abilities of the players in the dugout.) We choose one spinner at random from the bowl and then spin it 20 times, which corresponds to the 20 plate appearances of the hitter. We then record the spinner we chose (the value of the OBP area p) and the number of spins in the On-Base region of the spinner, which corresponds to the number of times on base for the hitter.

We repeat this process (randomly choosing a spinner and spinning it 20 times) for a large number of simulations—1000 of them, to be exact. Since we are interested in what we learn from 8 on base in 20 PAs, we focus on only the simulations where the spinner landed at On-Base 8 times for an observed OBP of

.400. We recorded the true OBPs (the ps) for these players, so we can ask the question: "What were the abilities of the players corresponding to these 8/20 spinner results?" Table 3-8 gives the results.

In our simulation, we observed the result "8 on base in 20 PAs" 97 times. Of these 97 occurrences, the hitter was a truly weak hitter (that is, with a .200 ability) 5 times, for a probability of 5/97 = .052. Since this is a small probability, we're pretty sure that this batter isn't a weak hitter.

Looking further at the table, we see that it is most likely that Mickey has a true OBP of .500, and it is almost as likely that he is a .400-OBP man—these two abilities have respective probabilities of .371 and .299. It is typical practice to group a few likely abilities that collectively have a large probability. Looking at the table, we see that these abilities:

$$p = .300, p = .400, \text{ and } p = .500$$

have a total probability of:

$$.175 + .299 + .371 = .845$$

So, based on observing 8 out of 20 PAs, we are pretty confident that Mickey has a true ability between .300 and .500. We call the interval [.3, .5] a 84.5 percent *confidence interval* for the unknown ability p.

What has Casey learned about the hitting ability of Mickey based on 20 plate appearances? Actually, very little. To say that a hitter's true OBP is between .300 and .500 doesn't say very much, since we observed from Chapter 1 that practically all of the OBPs of regular major league players fall between .300 and .500.

To emphasize how little is learned from 20 plate appearances, let's modify our example to include a more narrowly defined set of ability categories. Suppose that the OBP abilities of the players in the dugout are in the range .300, .310, .320, . . ., .600, and again the manager has no clue which players are good or poor, and the dugout contains an equal number of players of each ability level. As before, Casey selects players at random to be in the lineup, and one particular

						Total
Ability (p)	0.2	0.3	0.4	0.5	0.6	
Name	Weak	Below-average	Above-average	Excellent	Superstar	
Number of Players	5	17	29	36	10	97
Proportion of Players	0.052	0.175	0.299	0.371	0.103	1

TABLE 3-8   Abilities of Players Who Had 8 out of 20 On-Base in the Simulation Experiment

player gets on base 8 times out of 20 PAs. We want to find the probability that this hitter has a .300 ability, a .310 ability, and so on.

Figure 3-5 displays a line graph of the probabilities that the hitter has each of the possible OBP abilities. We see that the chance that the batter has a .300 ability is about .03, or 3 percent. In fact, each of the abilities between .300 and .500 has probabilities between 3 and 5 percent. So we can't make fine distinctions in ability—say, between a hitter with $p = .400$ and another hitter with $p = .410$—on the basis of an observed OBP of .400 for 20 PAs.

At this point, you may be thinking, "Okay, you've convinced me that you don't learn much from 20 PAs, but you must know about a player's ability on the basis of his season statistics. That's a lot more plate appearances."

To address this question, we repeat the above exercise using more data. Again we have a dugout full of hitters with abilities in the group [.300, .310, . . ., .600], and the manager has no clue which hitters have which abilities. A certain player (like Alomar) then plays an entire season and gets an OBP of .430 based on 682 PAs. We did a simulation like the one described above; Table 3-9 shows the probabilities of hitter ability given an OBP of .430 in 682 plate appearances.

Looking at the table results, we see that it is most likely that a hitter with an observed OBP of .430 actually is a true .430 OBP hitter. But other ability values

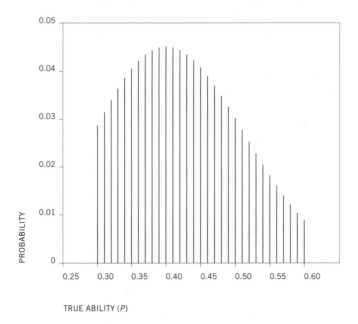

TRUE ABILITY (*P*)

FIGURE 3-5   Probabilities of true OBPs if a batter gets on base 8 times in 20 PAs.

							Total
Ability *(p)*	0.380	0.390	0.400	0.410	0.420	0.430	
Number of Players	2	11	22	33	64	65	
Proportion of Players	0.006	0.034	0.068	0.102	0.198	0.201	
Ability *(p)*	0.440	0.450	0.460	0.470	0.480	0.490	
Number of Players	47	44	22	8	3	2	323
Proportion of Players	0.146	0.136	0.068	0.025	0.009	0.006	1

TABLE 3-9   Abilities of Players Who Have a .430 OBP in 682 PAs

close to .430 are also very possible. If we group the most likely ability values, we see the following values:

$$.390, .400, .410, .420, .430, .440, .450, .450, .460$$

have total probability of:

$$.034 + .068 + .102 + .198 + .201 + .146 + .136 + .068 = .953$$

So this player's true OBP value is very likely to be in the [.390, .460] range.

Estimating Batting Ability Using a Confidence Interval

A basic task of professional statisticians is to provide bounds on how well an *estimate* (such as a season OBP) correctly identifies a *parameter* (in this case p, the probability of getting on base). A common approach to this inference problem is to calculate a confidence interval for the probability p, which we illustrated in the previous section. Using a standard formula taught in all introductory statistics classes, a 95 percent confidence interval for a probability has the following form:

$$\text{Estimate} \pm \text{a margin of error}$$

where

$$\text{Margin of error} = 1.96 \times \sqrt{\frac{\text{estimate} \times (1 - \text{estimate})}{\text{sample size}}}$$

To illustrate, suppose we are interested in learning about Alomar's true ability to get on base, which is measured by the on-base probability p. Our guess at Alomar's on-base probability is the observed season OBP value .422, which is based on a sample size of 682 plate appearances. So the margin of error is equal to

$$\text{Margin of error} = 1.96 \times \sqrt{\frac{.422 \times (1 - .422)}{682}} = .037$$

So a 95-percent confidence interval for Alomar's true ability to get on base p is

$$.422 \pm .037$$

or

$$[.385, .459]$$

This means that there is an excellent chance (a 95-percent chance to be precise) that Alomar's on-base probability p is between .383 and .457. (This interval is essentially the same interval that we computed in the previous section using a different rationale.)

This confidence interval gives an idea of how well we know the true ability of getting on base. After 682 plate appearances, we're pretty sure that Alomar's on-base probability is not .350 or .500 or any other value not in the confidence interval. However, values of p such as .390, .420, or .450 are all plausible, since they do fall within the interval.

Suppose it is May 1, 1999, and Alomar has only played one month in the season. He currently has an observed OBP of .422 based on 150 plate appearances. What have we learned about his on-base percentage p? Using the same formula, but with a sample size of 150 plate appearances instead of 682, the 95-percent confidence interval is as follows:

$$[.341, .499]$$

Since our confidence interval is pretty wide (about 150 points), we are pretty unsure about Alomar's true OBP based on one month of hitting data. Maybe he is average at getting on base and his true OBP probability is a mediocre $p = .341$, or maybe he is great in getting on base with a high OBP probability of $p = .499$. Both values of Alomar's true OBP are reasonable given only 150 plate appearances.

Of course, as plate appearances accumulate over time, we learn more about Alomar's true on-base probability. Figure 3-6 shows how an OBP confidence interval moves closer to the true on-base probability p as the number of plate appearances increases. The vertical lines in the figure mark different 95-percent confidence intervals around the observed OBP = .422 estimate for different number of PAs. This confidence interval rapidly narrows to (.35, .5) as the number of PAs increases to 100. After this point, the lengths of the confidence intervals decrease slowly. Even after 350 PAs (about half a baseball season for a

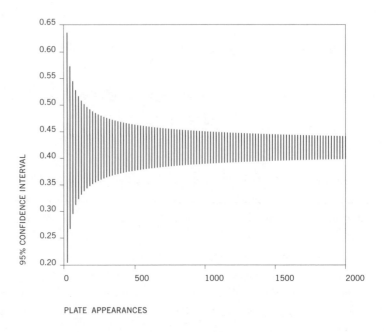

FIGURE 3-6   Graph of 95-percent confidence intervals for an on-base probability p as the number of plate appearances changes from 20 to 2000.

player starting in every game), the confidence interval has only narrowed to [.369, .471]. The final half-year provides little additional information; we saw that after Alomar's 682 PAs, the confidence interval has only narrowed to [.385, .459]. Even after a full season, we are not very sure of a player's true ability to get on base. Needless to say, this lack of precision in the observed OBP as an estimator of the probability of getting on base is never discussed by sportscasters or sportswriters.

Comparing Hitters

What does this mean? Basically, that seasonal on-base percentages are highly variable. A player with a .400 OBP for a season has a reasonable chance of having greater batting ability than a player who had an OBP of .410 in the same season. It *is* a fact that the .410 player had a better OBP than the .400 OBP player over the course of that particular season, but the .400 batter may still be the better batter—that is, have the higher probability of getting on base.

Let's consider a batter who at some point in the middle of a season has a .400 OBP. There is no doubt that up to this point the batter has done well. But from

the discussion we just had about the confidence interval for a true on-base percentage, we might question just how good this player really is. Is the .400 OBP a
true reflection of his ability, or is it just the result of good fortune? Clearly, again
from the earlier discussion, a lot depends on the number of plate appearances
represented by the .400 OBP. In particular, we might ask whether the batter is
truly better than other players with lower season OBPs. Let us consider three
other batters with the same number of plate appearances as the .400 batter.

Consider two batters—Joe has an observed OBP of .400 and Mike has an
observed OBP of .375 at a particular time during the season. Baseball people
will say that Joe is the better hitter simply because he is currently performing
better in getting on base. But what is the chance that Joe actually has a larger
true OBP, or larger ability to get on base than Mike? Figure 3-7 (bottom curve)
displays the probability that Joe (who is currently hitting .400) is truly better
than Mike (who is hitting .375) for a wide range of plate appearances. If both hitters have only 100 plate appearances, then the probability that Joe is better is
only about 65 percent, and if the players each have 600 PAs, then the probability Joe is better is still only 80 percent. We see that we are *not confident* that Joe
(who hits for .400) is better than Mike (who hits for .375) based on one season of
hitting data.

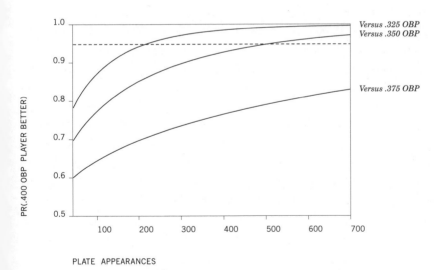

FIGURE 3-7   Graph of the probability that a player with an observed OBP .400 has a better
ability than players with observed OBPs of .325, .350, and .375 as the number of plate appearances increases from 50 to 700.

Figure 3-7 also compares our batter who hits .400 (Joe) with batters who have hit for .325 and .350 OBPs. The middle curve is a graph of the probability that Joe actually is superior to a guy who bats .350 based on different numbers of PAs, and the top curve displays the probability that Joe is superior to a hitter who bats .325. As expected, it is more likely that Joe has greater ability as the opponent's observed OBP decreases. Let's say that we are confident that Joe is better than the other batter if the probability he is better is 95 percent or higher. (This value is indicated by a dashed line in the figure.) Looking at Figure 3-7, we see that it takes 550–600 plate appearances (a full season of PAs) to say that the hitter who bats .400 is better than the hitter who bats .350. It takes fewer PAs to distinguish hitters who bat .400 and .325. From Figure 3-7, we see that we are confident the hitter who bats .400 is better than a hitter who bats .325 after 250 PAs, or roughly half a season.

As a baseball season progresses, we will be comparing the quality of hitters by means of statistics like the OBP. Halfway through the season, we are pretty confident that two hitters that have season OBP values 75 points apart (like .400 and .325) have different abilities to get on base, and that these relative abilities are reflected in their numbers. After an entire season, we can make finer distinctions between hitters, and we can say (with 95 percent confidence) that a hitter who has a 50-point observed OBP advantage (like .400 and .350) is the better hitter. It is hard, however, to make reliable distinctions when the margin of difference is less than 50 points, and we're not confident that a difference of 25 points between two hitters (like .400 and .375) is meaningful at all.

As you can see, statisticians aren't just people who make inferences. We recognize that the statistics that we observe over a season are only measures of the players' performance, and we use probability models to help learn about the players' abilities. Suppose some hot rookie, lets call him Max Marvelous, bats .350 next year. A typical baseball fan may conclude that Max is a great hitter. A statistician would come to a different conclusion: Max may be a great hitter based on this great hitting season, but we won't be convinced that Max has great batting ability until he maintains this great hitting performance for a number of seasons.

SITUATIONAL EFFECTS

chapter 4

It's April 4, 1999, and we're watching the ESPN broadcast of the opening game of the season, between the Padres and the Rockies. As each player comes to bat, the announcers give the viewers some insight on how the players perform in different situations. In this particular broadcast, we learn the following:

- Tony Gwynn is a very tough hitter with runners in scoring position.

- Vinny Castilla hits 100 points better at home than on the road.

- Todd Helton had a slow start in the previous (1998) season.

- Wally Joyner is pretty valuable to his team, since the Padres generally don't win when he is not in the lineup.

All of these statements rely on *situational statistics*—that is, they tell us, "in this situation . . . ," or, "under these circumstances . . . ," this or that tends to be true. In the last few years, we've found these kinds of statistics everywhere—in newspapers, on talk radio, from the broadcast booth during the game. Another example: a day after the ESPN broadcast, one of us, reading the newspaper (*The Findlay Courier*, April 5, 1999), encountered the following headline: "Indians Hope to Improve Their Situational Hitting." Reading the article, it seems that the Indians didn't perform well in 1998 in moving a runner along with a bunt, or getting a runner home from third with a hit. To improve on this, the team has done more in spring training to promote situational hitting, although leadoff hitter Kenny Lofton comments that failure to move a runner is sometimes just bad luck: "The one thing the stat don't show is how many times you hit the ball hard in those situations and having nothing to show for it. Maybe the ball was caught. Maybe the ball went foul. I know I had that kind of luck a lot last year." The article then says that Lofton hit a respectable .289 with runners in scoring position in 1998, but with runners on third and fewer than two outs, he hit .222 (4 for 18). Is Lofton just making excuses for himself, or does he have a point?

Surveying the Situation

In this chapter, we try to understand what we learn from situational statistics, and what we don't. We will focus on one basic measure, a player's batting average, and explore what can be learned by breaking it down into situational subsets: What was his average with men in scoring position, or on the road, or before the All-Star break? We'll take our data from the book *Player Profiles*, published by STATS, Inc., which contains one of the most extensive collections of situational statistics.

To start, we'll narrow down our discussion to one player, Scott Rolen, a third baseman for the Philadelphia Phillies. The 1998 season was Rolen's sophomore year in the majors, in which he attempted to show that his Rookie-of-the-Year performance in 1997 was not a fluke. See Table 4-1 for his stats.

First, we look at Scott's overall hitting. In 1998, he got 174 hits out of 601 at-bats, for a batting average of 174/601 = .290. We'll see later that this is a pretty good average—it's better than the batting average of a typical MLB regular player in 1998.

The table then breaks down these batting stats by a number of different situations. The vs. Right and vs. Left rows of the table show how Scott performed against right- and left-handed pitchers. Generally, it is believed that one hits better against pitchers of the opposite arm: Since Scott is a right-handed batter, one expects him to hit better against left-handed pitchers. Looking again at Table 4-1, we're a little surprised—Scott hit .292 against right-handers and .280 against left-handers.

Next, the table breaks down Scott's hitting by the type of pitcher faced. Some are classified as ground-ball pitchers, since their pitching tends to induce a lot of ground balls; others are characterized as fly-ball pitchers. We see that Scott did somewhat better against the fly-ball pitchers (.293) than the ground-ball (.276).

The next three situations break down the hitting data by the location of the ball park (home and away), the time of the game (day and night), and the playing surface (grass and turf, meaning artificial turf). Generally, ballplayers are thought to play better in their home ball parks than in opponents'. There are a

1998 SEASON

	AVG	AB	H
Season	.290	601	174
vs. Left	.280	132	37
vs. Right	.292	469	137
Groundball	.276	170	47
Flyball	.293	116	34
Home	.322	286	92
Away	.260	315	82
Day	.297	172	51
Night	.287	429	123
Grass	.254	232	59
Turf	.312	369	115
Pre-All Star	.303	333	101
Post-All Star	.272	268	73
Scoring Posn	.294	170	50
Close & Late	.279	111	31
None on/out	.287	115	33

	AVG	AB	H
First Pitch	.400	80	32
Ahead in Count	.348	135	47
Behind in Count	.205	249	51
Two Strikes	.225	298	67
Batting #3	.289	568	164
Batting #4	.313	32	10
Other	0	1	0
March/April	.271	96	26
May	.345	113	39
June	.295	105	31
July	.273	99	27
August	.245	106	26
Sept/Oct	.305	82	25
vs. AL	.311	61	19
vs. NL	.287	540	155

TABLE 4-1   Scott Rolen's Situational Statistics in 1998

number of reasons for this—players are more familiar with the characteristics of their own park, they are better rested since they aren't traveling, and they are being cheered by their fans. We see that Scott hit much better at home (.322) than at away games (.260), which is what we would expect. There doesn't appear to be much of a time-of-day effect—Scott hit just a little better during day games (.297) than during night games (.287). Also, he hit .312 on games played on artificial turf, compared with .254 on grass.

After noting that Scott hit better at home, the better average on turf should not surprise us. Scott's home park is Veterans Stadium in Philadelphia, which has artificial turf. So if we combine the situations "surface" and "home/away," we see that there really are only three situations—home, away on turf fields, and away on grass fields. One can figure out from the table that Scott hit .322 (92 for

286) on home-turf, .254 (59 for 232) on away-grass, and .277 (23 for 83) on away-turf. Generally, in analyzing situational data, one has to watch for situations or categories that are highly related or overlap. It is hard to tell if Scott really is a better hitter on turf since he played most of his turf games at home, in Philadelphia.

Returning to Table 4-1, the hitting data is also divided by different periods of the season. Scott hit .303 during the first half (before the All-Star Game) and only .272 during the second half. The month-by-month breakdown shows he had an especially hot May, batting .342, and a cool August, batting only .245.

The last grouping on the left-hand side of the table breaks down the hitting data by game situation. The heading "Scoring Position" indicates the player comes to bat with a runner at either second or third base. "Close and Late" occurs when the game is in the seventh inning or later and the batting team is leading by one run, is tied, or has the potential tying run on base, at bat, or on deck. "None On/Out" means that the player comes to bat with no runners on base and no outs. The "Scoring Position" and "Close and Late" situations represent times in the game where it is especially important for the player to get a base hit. In contrast, there is less pressure on a player when there are no outs and no runners on base. Here Scott hit for about the same average in all three situations—there is little evidence that he hits better or worse in pressure-packed situations.

The first grouping on the right-hand side of the table tells us how well Scott hit on different pitch counts. We see from the table that in plate appearances where Scott hit on the "First Pitch," he hit .400. In plate appearances where he was "Ahead in the Count" (where the number of balls exceeds the number of strikes), he hit for an average of .348. For "Behind in the Count" situations (pitch counts of 0–1, 0–2, 1–2, 2–2), he hit .205, and he hit for an average of .225 when there were 2 strikes in the count.

Last, the table tells us how Scott hit when he was batting number 3 and number 4 in the Phillies lineup. He played most of his games batting third, so this breakdown is not very interesting. Likewise, Scott's hitting performance against National League and American League teams is not that informative, since he played almost all of his games against NL teams.

Looking for Real Effects

It's pretty obvious that *Player Profiles* is great reading. A Phillies fan who especially likes Scott Rolen will have fun analyzing his breakdown statistics. Although Scott appears on the surface to be a .290 hitter, it seems his batting average during the season varied greatly, depending on particular circumstances. In the following situations, he appears to be an excellent (over-.300) hitter:

- in games played at home
- in games played on artificial turf (but we noted that this might be the same as home)
- in May

He appears to be especially hot when he swings on the first pitch (.400) and when he is ahead in the count (.348).

When we look at *Player Profiles*, we'll find a wealth of intriguing numbers. But do all of these numbers—these high and low situational batting averages—mean anything? In other words, do the observed differences between the averages in distinct situations correspond to "real" effects? Was Scott Rolen really a better hitter during home games? During the next season, should the Phillies manager bench Scott for games played on grass because of his sub-par 1998 batting average on that surface? Should we be surprised when Scott hits for a .400 average on the first pitch?

The preface of the *Player Profiles* book states:

> Not all .300 hitters are created equal. Last year, one hit .446 in April, and one hit under .190 that month. One hit over .400 after the All-Star Break and one hit under .220 for the final two months of the season. . . . When you think about it, calling a player "a .300 hitter" really doesn't say very much.

In the rest of this chapter, we'll use probability models to see if there really *are* differences between .300 hitters. Remember Kenny Lofton's comment about his bad luck hitting with runners in scoring position? We will try to explain how much of the variation in the situational data is due to good or bad luck. Specifically, we'll explain what we mean by "real" or "true" situational effects, and see what we learn about them from the situational data for all of the regular players in the 1998 season.

Observed and True Batting Averages

Recall our discussion about a *true proportion* and an *observed proportion* in Chapter 3: if we toss a fair coin in the air, we know that the chance of tossing heads is .5—this number represents the true proportion of heads. But if we toss the coin 20 times, we may get 8 heads for an observed proportion 8/20, or .4. As we stated, in any situation involving chance, it is likely that an observed proportion will be different from the true proportion.

In a similar fashion, we can define the concepts of observed and true batting averages. In 1998, Scott Rolen came to bat 601 times and got 174 hits, so we compute his batting average as follows:

$$174/601 = .290$$

This is his observed batting average, based on 601 opportunities to get a hit.

We measure Scott's ability to hit this season by a number p, which we call his true batting average. Like the coin probability of 50 percent, this number represents Scott's chance of getting a base hit in a single at-bat. It is very unlikely that his true batting average is equal to his 1998 observed batting average of .290.

Although Scott's true average p is unknown, we learn something about his true average by his performance during the 1998 season. Since he hit .290 for his 601 at-bats, we would guess that his true average p is close to .290. Actually, using the formula presented in Chapter 3, we can construct a 95 percent confidence interval for the true average. It turns out to be [.254, .326]. So we are pretty confident that Scott's hitting ability p is between .254 and .326.

Let's illustrate the difference between Scott's hitting ability and his season performance by means of a simple simulation. (A similar simulation experiment was performed for Roberto Alomar's on-base percentage in Chapter 3.) Suppose Scott is really a .300 hitter, and the chance p that he gets a hit in a plate appearance is 30 percent. Imagine a spinner, as illustrated in Figure 4-1, where the pointer can land anywhere on the circle. (Since we're assuming his hit probability is .3, the Hit region is 30 percent of the total area.) We simulate Scott's hitting results for a season by spinning the spinner 601 times. The total number of pointers that fall in the Hit region will be his number of hits for the season.

On the computer, we did this simulation 100 times, obtaining the number of hits for each of the 100 seasons (assuming that his true batting average is .300). Table 4-2 shows the number of hits that we observed. These hit numbers are converted to batting averages by dividing each by the number of at-bats (601). These 100-season batting averages are displayed using a stemplot in Figure 4-2.

We are assuming that Scott is truly a .300 hitter and remains a .300 hitter for 100 seasons. But the stemplot illustrates that this good hitter has a variety of good, mediocre, and bad seasons. In two unlucky seasons, he hit only .260. At the other extreme, in one season he was very fortunate and hit .346. These simulated results again demonstrate that a player's true batting average can be different from his season batting average. Also, the differences are substantial: we see an 86-point differential between Scott's best (.346) and worst (.260) seasons in these simulated seasons.

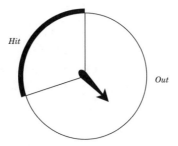

FIGURE 4-1   Spinner for simulating hitting for Scott Rolen.

179	172	177	198	178	176	194	184	172	180
177	166	180	190	184	181	165	185	168	170
192	183	185	190	169	191	169	189	188	179
203	187	203	185	171	190	177	185	175	191
183	190	181	175	171	176	173	177	162	156
172	156	183	181	176	170	171	173	165	182
190	176	189	192	177	176	197	195	174	173
174	169	208	160	180	172	176	187	178	167
174	196	171	184	190	182	168	188	173	191
168	189	187	172	206	174	198	201	177	182

TABLE 4-2   Number of Hits for 100 Seasons Assuming Scott Rolen Is a .300 Hitter

26	006
27	05568
28	0001113355556666668888
29	00001133333355555556688
30	0001113334446668888
31	11133444666666688899
32	346899
33	488
34	36

FIGURE 4-2   Stemplot of 100-season batting averages for a true .300 hitter.

Batting Averages of the 1998 Regulars

We focused on one hitter and noticed that there can be a significant difference between one's ability—that is, his true batting average—and his season batting performance. But what if we look at the batting averages of all the players in 1998? Can we use this bigger collection of data to make some conclusion about the true batting averages of all of these players?

In 1998, there were 246 players (in both leagues) who had at least 300 at-bats—we'll call these the "regular players," since 300 is about half the number of at-bats of a player who plays every game. In Figure 4-3, we've constructed a stemplot of the observed 1998 batting averages for these 246 players.

What do we see from this stemplot? The batting averages are approximately bell-shaped, and most of the averages are clustered in the .250–.299 range. The median batting average, that is, the value which divides the data set into a lower half and an upper half, is .276. The weakest hitter was John Flaherty, who had a .207 average. Two hitters stand out at the high end—John Olerud at .353 and Larry Walker at .363.

```
20 | 789
21 | 4788
22 | 11245
23 | 1222334566777789
24 | 01233455567889999
25 | 00111222233344455555667788999
26 | 00122344455566666667778888889
27 | 000111111222223334555556677788888888999999
28 | 000112222223344444555678899
29 | 00001112234444555556666889
30 | 000223445567788999
31 | 000112445556789
32 | 0111124457778
33 | 068
34 |
35 | 3
36 | 3
```

FIGURE 4-3   Stemplot of batting averages of 1998 regular players.

Two Models for Batting Averages

From this graph of the observed batting averages, what can we say about the true batting averages of these regular ballplayers? Remember we don't actually know the true hitting probabilities of these 246 players. All we know is their batting performance in the 1998 season. We will suggest two models for the true hitting probabilities for these players and see how well data that is simulated from these models mimic the actual 1998 batting averages.

A .276 Spinner Model

One simple model for hitting is based on the use of a random spinner (shown in Figure 4-4) similar to what was used for Scott Rolen. It consists of two regions (indicated by the heavy and light sections of the circle) that we'll call, respectively, "Hit" and "Out." The area of the Hit region corresponds to the true hitting probability of the player. Suppose that the player is average in ability—we'll call him "Joe Average." If Joe is a typical regular batter, it is reasonable (using 1998 hitting statistics) to let him have a .276 chance of getting a hit—this corresponds to a Hit area of .276. We call this model "a .276 spinner."

If Joe has 500 at-bats during the season, we can simulate a season of hitting for this average player by spinning our .276 spinner 500 times. We performed this simulation once, and the spinner landed 152 times on the Hits region, which corresponds to 152 hits during the season. Joe's batting average for this simulated season is then, 152/500 = .304. Note that the observed batting average for Joe is different from his true hitting probability. As we saw in the case of Scott

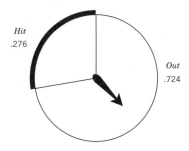

FIGURE 4-4   Spinner for an average (.276) hitter.

Rolen, that will typically be the case. On a computer we had Joe play 10 simulated seasons—Table 4-3 gives the number of hits and batting average for each.

We see that there is a lot of variation in Joe Average's season performance. In season 6, he was "hot" and batted .310. In contrast, he "slumped" in season 9 and batted only .236. Remember, Joe is always an average-ability hitter. Just by chance variation, he is having good and poor batting seasons. (In contrast, a sportswriter who observes Joe's .236 season would probably offer numerous explanations for his poor year.)

Do All Players Have the Same Ability?

The variability of the season batting averages for this average player raises an interesting question. Is it possible that *all* players have the same ability as Joe Average, and that the differences observed in player averages for the 1998 season are therefore just the result of random variability? This seems like a silly question—we think players have different hitting abilities—but it might be helpful to check this scenario.

Remember that a hitter of average ability is represented by a spinner with a Hit area equal to .276—a so-called .276 spinner. If all 246 players in the major leagues are all average hitters (that is, a set of Joe Average clones), then we have a set of 246 spinners, each spinner having a Hit area of .276.

To simulate a season of hitting for all 246 players, we just spin these 246 spinners many times, recording the number of Hits and Outs. After this simulated season is completed, we compute the season batting averages for all players, graph the simulated averages, and compare the results with the graph of the actual 1998 season batting averages.

Figure 4.5 shows what happened when we did this simulation for one baseball season. The actual 1998 batting averages are displayed in the left boxplot, and the simulated batting averages are displayed in the right boxplot.

SIMULATION

	1	2	3	4	5	6	7	8	9	10
HITS	152	121	131	147	126	155	126	128	118	144
ABS	500	500	500	500	500	500	500	500	500	500
AVG	0.304	0.242	0.262	0.294	0.252	0.310	0.252	0.256	0.236	0.288

TABLE 4-3   Simulated Data from 10 Seasons of Hitting by Joe Average

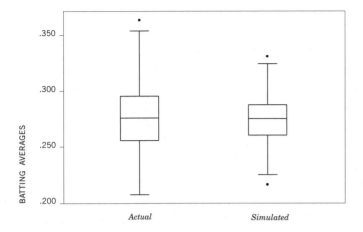

FIGURE 4-5   Boxplots of actual 1998 batting averages and simulated seasonal averages from one random spinner model.

What do we see? Both groups of batting averages are centered about the median value .276. But the actual 1998 batting averages are more spread out than the simulated averages. In the simulated season, only a couple of players hit less than .240 or more than .320. In contrast, there were a number of players in the actual 1998 season who had averages lower than .240 or higher than .320. This suggests that the "equal ability" model doesn't provide a good fit to the data.

When we repeated this simulation of baseball seasons many times, the result was the same. The actual 1998 season batting averages always had greater spread than the season batting averages simulated from the "Joe Average clones" model. What do we conclude? A .276 spinner model does *not* work for baseball hitting data, which means that hitters *do* have different abilities. A more complicated model is needed to represent baseball hitting data. As we said before, this is a pretty obvious conclusion—it would seem ludicrous to say that players all have the same hitting ability—but it illustrates the basic method we'll use to check the suitability of other models.

A Model Using a Set of Random Spinners

From our investigation of the .276 spinner, we concluded that players have different hitting abilities. The next question is, how can we represent these different abilities? When we take physical measurements of the general population and collect data about different characteristics—such as height, arm span, foot size, and so forth—we find that most of the measurements cluster in the middle and attenuate (or "thin out") at the high and low extremes. We call this data distribution "normal," and when we put it into a graph, we get the familiar bell-shaped curve. This normal curve is also found in data describing the abilities of people. For example, if you give 100 people a standardized test on some subject matter such as math, the test scores will be approximately bell-shaped.

This bell-shaped curve is also useful for describing the hitting abilities of ball players. Suppose that the true hitting proportions of regular Major League ball players are described by a normal curve. We'll let the center of the curve be .276, which is the typical season batting average for 1998. Next we have to decide on the spread of this curve. We know that ball players have different abilities, and the spread of this curve will tell us how different the abilities can be. We choose a spread for the curve so that the season batting averages that are simulated resemble the actual batting averages for the 1998 season. How we figure out this spread is a bit complicated. But it turns out that if we let the normal curve have a standard deviation of .021, the season batting averages are a pretty good match to the actual 1998 data.

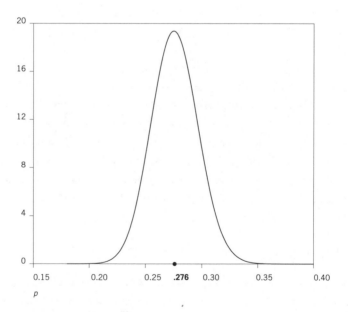

FIGURE 4-6   Normal model for true hitting probabilities.

The technical name for this model is a *random-effects model*, or a random-ability model. To better understand this model, let's describe in some detail how one could simulate hitting of all the regular players in a baseball season.

- The first thing we do is to *choose the different abilities* of the 246 players who are playing the 1998 season. We choose these abilities randomly based on the normal curve in Figure 4-6. We did this on a computer and got the true baseball averages shown in Table 4-4.

- To see where these true batting averages come from, we've graphed these 246 averages in Figure 4-7 and placed the normal-ability curve on top. (This figure uses a *dotplot,* where each batting average is represented by a dot on a number line.) We see that most of the true averages are in the 250–300 range, which is what we predict from the normal curve.

278	**259**	**290**	281	281	270	273	249	241	271	284	270
226	**297**	**301**	261	282	265	248	299	261	276	276	241
310	**223**	**252**	254	281	250	269	256	251	255	268	280
244	276	277	283	286	303	264	281	276	264	321	271
246	313	283	253	289	303	257	279	273	252	301	276
287	261	262	283	278	315	270	322	308	235	241	264
272	276	294	261	261	272	276	282	298	289	239	291
293	289	304	283	262	273	225	286	278	264	262	253
275	284	269	266	275	272	256	303	285	303	266	253
293	277	260	274	234	299	255	262	304	257	267	265
310	278	253	252	312	317	310	250	272	272	282	264
255	267	299	326	281	270	291	266	315	299	250	262
304	284	284	240	281	290	263	255	272	254	274	282
305	280	265	293	281	290	265	294	282	289	254	274
308	285	236	286	303	289	305	304	257	228	314	284
276	267	244	281	247	258	272	292	284	248	307	277
315	251	260	260	274	255	256	226	258	281	272	272
274	280	265	261	287	267	281	256	273	292	257	279
271	274	268	232	273	305	290	268	262	281	268	265
277	302	288	255	296	225	271	277	267	272	244	298
259	284	264	308	279	271						

TABLE 4-4 Simulated True Batting Averages for 246 Players

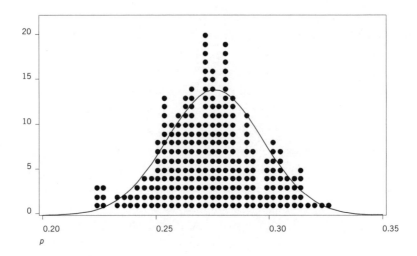

FIGURE 4-7   Dotplot of the true batting averages with a normal curve placed on top.

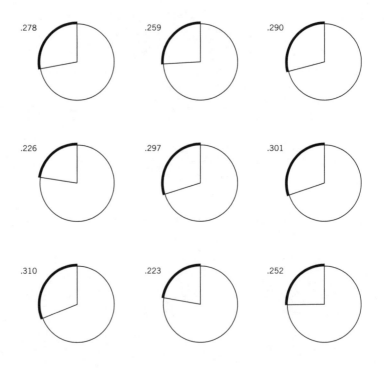

FIGURE 4-8   Spinner models for nine hitters from the random-ability model.

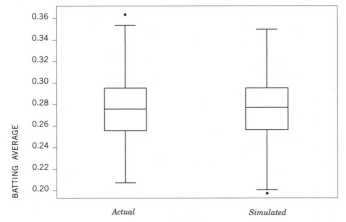

FIGURE 4-9   Boxplots of actual 1998 batting averages and simulated seasonal averages from the many spinners model.

- Based on the 246 true batting averages we have simulated, we construct 246 computerized spinners. Each spinner corresponds to a particular player, and the Hit area for his spinner will be equal to the true batting average for that particular player. So if Scott Rolen is assigned a .308 true average, his spinner would have a relatively large Hit area, and Albert Belle, with a .251 true average, would have a spinner with a smaller Hit area, and so on. The nine spinners shown in Figure 4-8 correspond to the boldface averages in the upper-left corner of Table 4-4.

- We then spin all of these spinners for a full season of hitting, where each player has the same number of at-bats as he actually had in the 1998 season.

We did this simulation for a single baseball season and computed the season batting averages for all 246 players. How did these simulated batting averages compare with the actual 1998 averages? Figure 4-9 compares the two sets of averages by means of two boxplots. Looking at the boxplots, we see that they seem to mimic the actual 1998 averages pretty well. For the actual 1998 data, there is one extreme average at the high end, and the simulated data has one unusually low average. In any event, this simulation confirms that the "many random spinners" model does a pretty good job of predicting the distribution of seasonal batting averages.

Situational Effects

What have we learned from looking at the 1998 batting averages?

- Batters appear to possess different hitting abilities.

- We can describe these different abilities by means of a group of true hitting averages. The group of true batting averages forms a bell-shaped curve centered on the typical value .276.

Now we're ready for a discussion of situational effects. How do players' batting averages change across different situations? Specifically, using the data from *Player Profiles*, we will compare players' batting averages in the following situations:

- during home games and away games
- against pitchers of the opposite arm and the same arm
- against ground-ball pitchers and fly-ball pitchers
- during day games and night games
- on games played on grass and turf
- on games played before and after the All-Star Game
- when the team is in scoring position with none on and none out
- when the pitcher is ahead in the count with two strikes

First, let's discuss what is commonly believed about the importance of several of these situations.

Home vs. Away

It's well known that all ball parks are not created equal. They differ in the distances from home plate to the fences, the size and shape of foul territory, the climate, and countless other particulars. And it is believed, by players and coaches and fans, that these differences have a significant impact on hitting. We hear that batters who regularly play in a park that is supposedly "friendly" benefit from that park. And it is widely accepted that players hit better at their home ball park than they do on the road. It's more comfortable to play in one's own park, goes the thinking: the players get to stay in their homes and drive themselves to work, while the away team which has to travel on planes and buses and stay in motels. And the home team, of course, is cheered on by local fans.[1] For all of these reasons, one expects players to have higher batting averages at home compared to away.

---

[1] Contrary to myth, even in our home town of Philadelphia.

Turf vs. Grass

It is well known that balls hit on artificial turf will behave differently from those hit on natural grass. It is believed that artificial turf will increase the number of doubles and triples, since line drives into the outfield will move fast and more likely evade the outfielders. Also, balls hit in the infield on artificial turf, we're told, will more likely reach the outfield. For these reasons, it is believed that "turf," as the artificial surface is called, has a positive impact on players' batting averages.

The Count

When a pitcher faces a batter, there are 12 possible pitch counts (0–0, 0–1, 0–2, 1–0, 1–1, 1–2, 2–0, 2–1, 2–2, 3–0, 3–1, 3–2). To an experienced fan, each count conveys a feeling about the batter's chances of getting a hit. For example, if the hitter is facing an 0–2 count, it is generally believed that the pitcher has a strong advantage and the hitter has a small chance of getting a hit. (Actually, the hitter is likely to strike out after an 0–2 count.) In contrast, a batter with a 3–0 count can be very relaxed and confident and has a high probability of walking or getting a hit. Thus it would seem that a player's ability to get a hit would vary greatly depending on the pitch count.

Opposite Arm vs. Same Arm

One of the fundamental managerial strategies is to have a hitter bat against a pitcher of the opposite arm. This strategy is based on the belief that it is easier to hit a pitch that's coming toward you than a pitch moving away from you. As Casey Stengel once said:

> There's not much to it. You put a right-hand hitter against a left-hand pitcher and a left-hand hitter against a right-hand pitcher, and on cloudy days you use a fastball pitcher.

According to this logic, one expects a player to have a better batting average against pitchers of the opposite arm.

Models for Situational Effects

To understand which of the above situational effects are "real," we'll use the same basic strategy that was used in analyzing the set of batting averages. We will propose a few basic models for situational effects, then we will fit these basic models to the 1998 batting-average data. Based on this fitting, we'll see which models seem to predict the pattern of effects we see in the 1998 data.

To describe these models, let's use a hypothetical situation that is *not* represented in *Player Profiles*. Suppose we break down the hitting data by the size of the crowd—*below average* (whatever "average" is for that particular stadium), and *above average*. We want to know if the size of the crowd has any effect on the players' batting averages. For ease of description, we will refer to these two scenarios as "small crowds" and "big crowds."

Suppose one of our favorite players, Tony Gwynn, has 80 hits in 250 at-bats on small-crowd days and 100 hits in 250 at-bats on big-crowd days. Given this information, we observe the situational batting averages in Table 4-5. Looking at the table, we might be tempted to say, "Wow!" Tony hit 80 points higher when the attendance is high–apparently, he really loves to play in front of big crowds!

But wait a minute. We are learning that there is a lot of variation in batting data for a single season. Maybe Tony doesn't care if he's hitting in front of a big or small crowd, and there actually isn't any *real* situational effect due to attendance. The attendance effect of 80 points that we see in the hypothetical season above might be due to chance variation. Tony may actually have the same true batting average on small- and big-crowd days, but by luck he just happened to do much better this year on big-crowd days. What we're trying to do here is to decide how much of the variation in situational hitting data is due to *real effects* and how much of the variation is due to luck or *chance variation*.

For each player, let's define two true batting averages. The first average p_B is the true batting average of the player when he plays in front of *big* crowds, and the second average p_S is the true average in front of *small* crowds. We'll call *the true situational effect* the difference between the two batting averages:

$$\text{True effect} = p_B - p_S$$

We want to learn about the sizes of the true effects for all of the hitters in 1998.

In the description of the models to follow, it will be convenient to write the two batting averages as

ATTENDANCE

	Small Crowd	Big Crowd	Difference
Batting Average	80/250 = .320	100/250 = .400	.400 – .320 = .080

TABLE 4-5   Hypothetical Situational Hitting Data for Tony Gwynn

True batting average for small crowds: p_S

True batting average for big crowds: $p_B = p_S + EFF$

Here we're using the abbreviation EFF to stand for the *True Situational Effect*. To put this another way, EFF measures how much better a player hits when there is a good crowd in the stands.

When we talk about situational effects, there are three possible scenarios. We'll describe these models using our hypothetical above-average/below-average attendance situation.

Scenario 1 (No Situational Effect)

One possibility is that there is *no true effect* due to attendance. A major-league ball player has been playing in front of crowds his whole life, and maybe he is oblivious to the size of the crowd. If this is a reasonable statement, then there is no reason to expect the batters to hit for a different average on small-crowd and big-crowd days.

If there is no situational effect due to attendance, what would true batting averages look like? We would see a distribution like the one shown in Figure 4-10, which shows the same normal curve we saw in Figure 4-6.

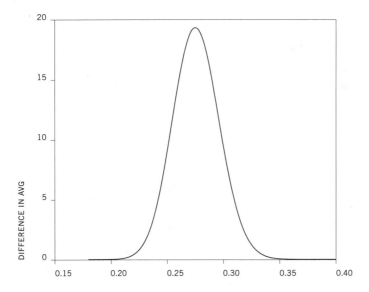

FIGURE 4-10   Normal curve for batting averages when there is no situational effect due to attendance.

Scenario 2 (Situational Bias)

Let's consider a different scenario. Suppose hitters actually like big crowds, and when they see a big crowd, they get excited and try harder. And since they try harder, they are generally more successful in hitting. If this is true, then a player's true batting average on big-crowd days will be higher than the player's true batting average on small-crowd days. Also, we assume here that the big crowd has the same effect on all of the players. So if this effect is, say, 40 points— that is, the average is 40 points higher for big crowds than for small crowds— then every player will have this same effect. The statistical term for this type of behavior is a *bias*.

Table 4-6 and Figure 4-11 illustrate what we mean by a bias. Suppose that the hitting abilities of the players *on small crowd days* follow a normal curve with mean .256 and standard deviation .021. In Table 4-6, we show the true batting averages for 50 representative hitters on poor-attendance days. In the top graph of Figure 4-11, we display these true averages using a dotplot.

Next, suppose that there is a bias of 40 points. That means that every player bats for a 40-point higher average when there is a big crowd, in which case:

$$EFF = .040$$

PLAYER   TRUE AVG

PLAYER	TRUE AVG	PLAYER	TRUE AVG	PLAYER	TRUE AVG	PLAYER	TRUE AVG
A	.258	N	.262	AA	.261	NN	.257
B	.239	O	.245	BB	.230	OO	.263
C	.270	P	.228	CC	.249	PP	.266
D	.206	Q	.279	DD	.236	QQ	.283
E	.277	R	.241	EE	.231	RR	.244
F	.261	S	.256	FF	.235	SS	.261
G	.261	T	.256	GG	.248	TT	.256
H	.281	U	.251	HH	.260	UU	.244
I	.250	V	.264	II	.254	VV	.301
J	.253	W	.250	JJ	.278	WW	.251
K	.229	X	.221	KK	.251	XX	.226
L	.221	Y	.234	LL	.224	YY	.293
M	.241			MM	.256		

TABLE 4-6   Fifty Representative True Batting Averages for Games with Small Crowds

To get the true batting averages for our 50 hitters *on big-crowd days*, we simply add 40 points to each *small-crowd* batting average in Table 4-6. The resulting averages are shown in Table 4-7.

The bottom graph of Figure 4-11 displays the true "big crowd" averages. Comparing the two dotplots, we see the effect of the situational bias. All of the dots in the top graph have been shifted to the right by 40 points to get the dots in the bottom graph. This emphasizes the fact that a situational bias means that all players have the same batting-average improvement due to a good park attendance.

Scenario 3 (Situational Effect Depends on Ability)

The third scenario is the most complicated description of what may be going on. Maybe there really *is* a boosting effect due to the size of the crowd, but the size of the effect depends on the player. For example, suppose that there are two players, Joe Cool and Harry Hyper, who react differently to big crowds. Joe is good in not letting outside influences affect his hitting. His true batting average when there are big crowds is approximately equal to his average when there is no one in the stands. Harry, in contrast, feeds on whatever energy level is present in the ball park. If attendance is low and no one is cheering, he is complacent, and his batting average suffers. On the other hand, if the ball park is filled to capacity

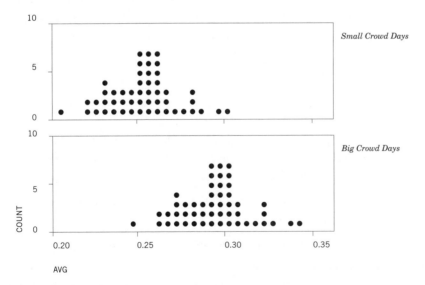

FIGURE 4-11   Dotplots of 50 typical "small crowd" and "big crowd" true batting averages when there is a situational bias of 40 points due to attendance.

PLAYER   TRUE AVG

A	.258+.040	N	.262+.040	AA	.261+.040	NN	.257+.040
B	.239+.040	O	.245+.040	BB	.230+.040	OO	.263+.040
C	.270+.040	P	.228+.040	CC	.249+.040	PP	.266+.040
D	.206+.040	Q	.279+.040	DD	.236+.040	QQ	.283+.040
E	.277+.040	R	.241+.040	EE	.231+.040	RR	.244+.040
F	.261+.040	S	.256+.040	FF	.235+.040	SS	.261+.040
G	.261+.040	T	.256+.040	GG	.248+.040	TT	.256+.040
H	.281+.040	U	.251+.040	HH	.260+.040	UU	.244+.040
I	.250+.040	V	.264+.040	II	.254+.040	VV	.301+.040
J	.253+.040	W	.250+.040	JJ	.278+.040	WW	.251+.040
K	.229+.040	X	.221+.040	KK	.251+.040	XX	.226+.040
L	.221+.040	Y	.234+.040	LL	.224+.040	YY	.293+.040
M	.241+.040			MM	.256+.040		

TABLE 4-7   Fifty Representative True Batting Averages for Games with Above-Average Attendance

and the crowd is cheering, Harry gets pumped and hits for a high batting average. Harry would have a large situational effect due to the park attendance.

In this scenario, we are saying that players actually possess different abilities to use the situation. Harry is more successful than Joe in using the crowd to his advantage. If Harry and Joe are both second baseman, then the manager might prefer to use Joe on small-crowd days and Harry on high-attendance days.

Finding Good Models

We've described three possible scenarios for a given situation effect. We'll nickname these scenarios as "no effect," "bias," and "ability effects."

- *No effect*. There is no true situational effect. Any differences that we observe in the season situational batting averages are solely due to chance variation.

- *Bias*. There is a true situational effect, but it is the same for all players.

- *Ability effects*. Players have different true situational effects.

How do statisticians find the best model for the 1998 hitting data? Actually, we don't want to bore you with a long explanation of the method of finding the model. What's more important here is an understanding what we mean by "best." We know that baseball hitting is a complicated process, and no model, including the ones described above, will perfectly describe what is going on with respect to situational effects. But what we do in statistics is try to find a single simple model that seems to explain pretty well the hitting data that we observe in *Player Profiles*. There are two important aspects of the model. First, we want the model to be simple so that it is understandable. Second, the model should be good in the sense that it makes reasonable predictions about current and future baseball data. The "many random spinners" model for batting averages is an example of a good statistical model. It is easy to understand, and it predicts, reasonably well, the pattern of season hitting data that we observe.

What Do Observed Situational Effects Look Like When There Is No Effect?

Before we look at the situational data of the 1998 hitters, it will be helpful to consider a scenario where there is *no* situational effect. We will simulate situational data for a season, and by looking at the observed situational batting averages, we'll understand the great variability that is inherent in this type of data.

Let's recall the "random spinners" model for hitting data. Here each of the 246 hitters has an associated random spinner, where the Hit region in the spinner corresponds to the true batting average. We spin these spinners for an entire season of hitting, and in this way simulate the season batting averages for the group of players.

We introduce a new situation where we know there is no effect. Let's suppose that, when we do this simulation, half of the time we spin the spinner in the *dark* and the other half of the time we spin in the *daylight*. So, for example, consider Roberto Alomar, who has 588 at-bats in the 1998 season. For 294 of the at-bats, we'll spin Alomar's spinner in the dark, and for the remaining 294 at-bats we'll use the spinner in the daylight. After we use the spinner for all 588 at-bats, we will record the number of hits that Alomar gets in the dark and in the light. We will compute Alomar 's batting average in each situation, and then we can compute the observed situational effect:

$$Observed\ situational\ effect = AVG_{dark} - AVG_{light}$$

When we did one simulation, Alomar got 87 hits in 294 AB in the dark for a batting average of 87/294 = .296. In the light, he had 71 hits in 294 AB for an average of .242. Alomar's observed situational effect in this case is:

$$Observed\ situational\ effect = .296 - .242 = 0.54$$

So Alomar hit for 54 more points in the dark in this simulation season. To many people, this would represent a significant effect—Alomar must like the dark! But of course this is not the case, since it was the spinning, and not real hitting, that took place with the lights out.

To investigate further, we repeated this simulation for all 246 players (using their 1998 at-bat totals) and computed situational effects for all players. In Figure 4-12, we graphed the 246 effects using a stemplot.

In this silly example, does there exist a true situational effect? If the same spinner is used in the dark and in the light, do you think the chance of getting the pointer to land in the Hit region will change depending on the light in the room? Of course not. If you spin the spinner the same way each time, the chance of getting a Hit will remain the same regardless of the light. We *know* in this case that there is no true situational effect.

Now look at the stemplot in Figure 4-12. Even though there is no dark-light effect, some of the players have large dark-light effects for this simulated season. One player batted .358 in the dark and .241 in the light for a whopping effect of .358 − .241 = .117. On the other side, there was one player who batted .262 in the dark and .372 in the light—110 points greater in the light. The situational effects are bell or normal shaped, centered about the average value of 0, which is what we would expect. But the spread of these effects is large, and practically all of the observed effects fall between −100 and +100 points. This simulation demonstrates that, even when there is nothing going on, the observed situational effects can look deceptively interesting.

```
-1 | 10
-0 | 99998888
-0 | 766666666
-0 | 55555555544444444444444
-0 | 33333333333333332222222222222222222
-0 | 11111111111110000000000000000
 0 | 000000000000000000000000000011111111111111111111111111
 0 | 22222222222222222222222222222233333333333333333333333
 0 | 44444444444444555555555
 0 | 666777
 0 | 8889
 1 | 0111
```

FIGURE 4-12 Stemplot of seasonal situational effects when there is not a true effect.

The Last Five Years' Data

Although we are focusing on interpreting the 1998 situational data, *Player Profiles* also gives the same situational hitting data for the previous five-year period, from 1994 through 1998. This five-year data is very useful for understanding the significance of observed situational effects. Looking at the book, we see that Butch Huskey in 1998 hit for an average of .299 against left-handed pitchers and only .230 against right-handers—here we observe a situational effect of 69 points (.299 − .230 = .069). Does this mean that Huskey (who is a right-handed hitter) is a much better hitter against lefties? Maybe or maybe not. It's possible that Huskey has the same hitting ability against lefties and righties, but by chance he just happened to do better against lefties in 1998. One way of checking if this 69-point differential is real is to look at his performance over the last five years. If he exhibits the same effect in the previous years, one would have more confidence that the situational effect really exists. The data in Table 4-8, taken from *Player Profiles,* helps us take just such a look at Huskey's five-year history.

We observe a left-right average difference of .301 − .254 = .047, which is pretty large. So we might conclude that Huskey is a much better batter against lefties But, wait: the last five years' data includes the year 1998. So one reason that Huskey has a large five-year effect is that he experienced a large effect in 1998. It would be better to remove the 1998 data from the last five years, creating data for the four-year period 1994–1997, as shown in Table 4-9.

We see that in the last four years, Huskey had a .301 − .263 = .038 average difference. This is reasonably large, so we believe that Huskey does hit for higher average against lefties. But the 1998 difference of .069 seems to be larger than his true advantage. We will use the last four years data to give support to our conclusions about the true situational effects.

LAST FIVE YEARS (1994–1998)

	AB	H	AVG
Left	389	117	0.301
Right	955	245	0.254

TABLE 4-8   Butch Huskey's Batting Data for the Five-Year Period 1994–1998

LAST FOUR YEARS (1994–1997)

	AB	H	AVG
Left	272	82	0.301
Right	703	185	0.263

TABLE 4-9   Butch Huskey's Batting Data for the Four-Year Period 1994–1997

```
-1 | 9                                             -1 | 6
-1 |                                               -1 |
-1 |                                               -1 | 3
-1 |                                               -1 | 00000
-1 | 00                                            -0 | 888
-0 | 98888                                         -0 | 777777666666666
-0 | 7777666666                                    -0 | 555555444444444444444
-0 | 5555555555555554444444444                     -0 | 33333333333333333333333322222222222222222222222
-0 | 33333333333333332222222222222                 -0 | 1111111111111111111111100000000000000000000000
-0 | 1111111111111111000000000000000000000000000    0 | 0000000000000000001111111111111111111111
 0 | 000000000000000000001111111111111111111111111   0 | 22222222222222222222222233333333333333
 0 | 2222222222222223333333333333333               0 | 444444445555555555
 0 | 4444444444444444455555555555                  0 | 6666666667777
 0 | 666666666677777                               0 | 889999
 0 | 88888888                                      1 | 0
 1 | 0001                                          1 | 23
```

FIGURE 4-13   Stemplot of observed 1998 differences AVG(pre-All-Star Game) – AVG(post-All-Star Game).

FIGURE 4-14   Stemplot of observed 1998 differences AVG(day games) – AVG(night games).

The "No Effect" Situations

Recall the simulation in which we spun our spinners half the time in the dark and half the time in the light. In that situation, even when we knew that there was no situational effect, the observed situational effects fell between –100 and +100 batting average points. Interestingly, we see this same pattern of season effects for all of the following situations:

- pre-All-Star Game vs. post-All-Star Game
- day games vs. night games
- grass vs. turf

In Figures 4-13, 4-14, and 4-15, we have used stemplots to graph the observed 1998 situational effects for pre/post, day/night and grass/turf for all 246 players. We see in this *actual data* basically the same pattern of effects that we saw in our light/day *simulated data*. For each of these three situations, the player effects are bell-shaped around 0 and spread out between –100 and +100 points.

To emphasize this point, Figure 4-16 displays parallel boxplots of the 1998 effects for the pre/post All-Star game, day/night, grass/turf situations together with the hypothetical dark/light values from our situation. Note the similarity of these four datasets, in terms of both the average value and the spread.

So we conclude that their is *no general effect* for these three situations. Players don't generally hit any better or worse in the last half of the season than the first half of the season. There is no general hitting advantage or disadvantage in playing a night game compared to a day game. And there is no general hitting effect with regard to the type of field (grass or turf).

One way to demonstrate the lack of a general effect for these situational variables is to compare the 1998 observational effect with the last-four-years (1994–1997) effect for all of the players. For example, consider the pre-All-Star/post-All-Star effect. Table 4-10 shows the pre-All-Star and post-All-Star batting averages for four of the players.

We see from Table 4-10 that Edgardo Alfonzo hit 21 points better in the second half of the season in 1998, but hit only 2 points better in the second half in the previous four years. Jermaine Allensworth hit 86 points better in the first half in 1998, but hit 1 point worse in the first half in 1994–1997.

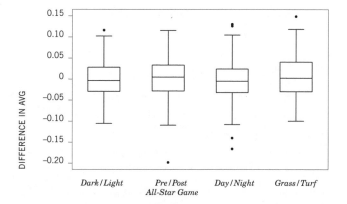

FIGURE 4-15 Stemplot of observed 1998 differences AVG(grass) – AVG(turf).

FIGURE 4-16 Boxplots of pre/post-All-Star game, day/night, and grass/turf 1998 situational effects. For comparison, a boxplot of the simulated dark/light effects is shown.

	1998			1994–1997		
PLAYER	PRE	POST	PRE–POST	PRE	POST	PRE–POST
Edgardo Alfonzo	0.267	0.288	–0.021	0.287	0.289	–0.002
Jermaine Allensworth	0.300	0.214	0.086	0.257	0.258	–0.001
Roberto Alomar	0.291	0.271	0.020	0.323	0.307	0.016
Sandy Alomar	0.261	0.196	0.065	0.315	0.271	0.044

TABLE 4-10   Batting Averages for Four Players Before and After the All-Star Game

Suppose that for each player we record the pre/post effect in 1998 and the pre/post effect in 1994–1997, and then graph these values in a scatterplot, as shown in Figure 4-17. We don't see any increasing or decreasing pattern in the graph, which means that there is little relationship between the players' pre/post situational effect in 1998 and the corresponding effects in the four previous years.

Since there is no general effect for these three situations, does this mean that there is no real situational effect for any of the individual players? No. It is possible that some players do take advantage of some situations. For example, it is possible that some batters have a stroke that is especially well-suited for artificial turf, and so they hit for a higher true average on turf than on grass. However, our analysis seems to indicate that this turf/grass effect, if it exists at all, applies only to a relatively small group of players. If many players had a turf/grass effect, then we would see it in the observed situational effects. But the data we see is very consistent with a model where there is no turf/grass situational effect for any of the players.

The "Bias" Situations

For three other situations, namely home vs. away, ground-ball vs. fly-ball, and same-handed vs. opposite-handed, there is evidence of a general situational effect. Generally, one can say that:

- Hitters bat 12 points better at home games compared with away games.
- Hitters bat 12 points better against ground-ball pitchers than fly-ball pitchers.
- Hitters bat 15 points better against pitchers of the opposite arm than pitchers of the same arm.

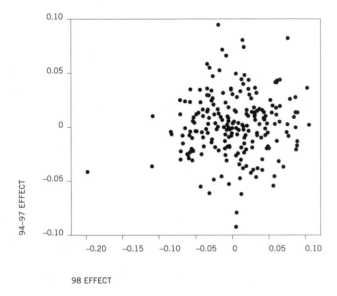

FIGURE 4-17   Scatterplot of 1998 situational effects and the previous four-year period situational effects for the pre/post All-Star Game situation.

Let's try to understand what this general effect means. It is well known that it is easier to hit in some ball parks (Coors Fields, in Denver, is an example) and relatively difficult to hit in others (Dodger Stadium comes to mind). So one would expect that players that hit in easy-to-hit or hard-to-hit ball parks might have different true home vs. away effects from other players who play in "average" ball parks. However, the ball-park effect is somewhat diluted, since players don't play all of their games in their home ball park, and all players have opportunities to hit in easy-to-hit or hard-to-hit ball parks.

What this general effect is telling us is that a home field appears to have the same impact on all of the players who regularly play in that ball park. Likewise, facing a ground-ball pitcher (instead of a fly-ball pitcher) has the same positive effect on all hitters, and facing a pitcher of the opposite arm has the same beneficial effect (15 points) on all hitters. Again, it should be emphasized that some players might really take advantage of the situation relative to other players. For example, one player might really take advantage of his home ball park and have a true home vs. away effect. But our analysis says that there are not too many players with unusually large or small situational effects, and the 1998 data is consistent with a model which says that the situation has the same impact on all players.

```
-0 | 98
-0 | 77777766666666
-0 | 55555444444444
-0 | 3333333333333322222222222222222222
-0 | 1111111111111111110000000000000
 0 | 00000000000000000000001111111111111111111111
 0 | 2222222222222222222222223333333333333333333333
 0 | 44444444444444444445555555555555
 0 | 6666666677777777
 0 | 889
 1 | 0000001
 1 |
 1 | 4
```

```
-0 | 9
-0 | 6666
-0 | 555544444
-0 | 33333333322222222222222222
-0 | 11111111111111111111100000000000
 0 | 000000000000000011111111111111111
 0 | 2222222222222222233333333333333333
 0 | 44444444444444444455555555555555
 0 | 6666666666666777777
 0 | 888899999
 1 | 00011
 1 | 2
 1 | 4
 1 | 7
```

```
-1 | 4
-1 | 3222
-1 | 00
-0 | 999888888
-0 | 7777666666666
-0 | 5555555555544444444444444444
-0 | 3333333333333322222222
-0 | 1111111111111100000000000
 0 | 000000000000001111111111111111111
 0 | 2222222222222223333333333333333333333333
 0 | 4444444445555555
 0 | 66666666667777777777777
 0 | 88888888888889999999999
 1 | 000011
 1 | 233
 1 | 55
 1 |
 1 | 8
 2 |
 2 |
 2 | 5
```

The stemplots of the observed situational effects for home/away, opposite/same, and ground-ball/fly-ball are shown in Figures 4-18, 4-19, and 4-20. The home/away and opposite-handed/same-handed stemplots resemble the day/night, pre/post, and grass/turf graphs that were shown earlier. The only difference is that the average home/away effect is about 12 points and the average opposite-handed/same-handed effect is about 15 points. The ground-ball/fly-ball effects are more spread out—they range between −140 and +250 points. There is a simple reason for this wider spread—the number of at-bats for these categories is small (not every pitcher is classified as a ground-ball or fly-ball type) and so there is more variation in the batting averages for these categories.

Figure 4-21 shows boxplots of the 1998 effects for these three situations and contrasts these effects with the simulated effects from our artificial dark/light example. Note that the average values of the home/away and opposite/same effects are a little larger than the dark/light effects, but the spreads of these three datasets are similar.

Since these three situations are biases, they affect all hitters the same way, and there is no reason that a player who has a large situational effect one season will also have a high situational effect the next

FIGURE 4-18   Stemplot of observed 1998 differences AVG(Home − Away Games).
FIGURE 4-19   Stemplot of observed 1998 differences AVG(opposite − same-handed).
FIGURE 4-20   Stemplot of observed 1998 differences AVG(ground-ball − fly-ball pitcher).

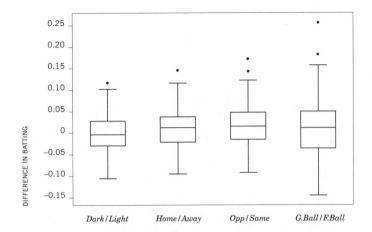

FIGURE 4-21   Boxplots of home/away, same-handed/opposite-handed, and ground-ball pitcher/fly-ball pitcher 1998 situational effects. For comparison, a boxplot of the simulated dark/light effects is also shown.

season. To see this, let's compare the same-handed/opposite-handed situational effects for the 1998 season and the 1994–1997 seasons. Suppose that, for each player, we record the same-handed/opposite-handed effect in 1998, and the same-handed/opposite-handed effect in the period 1994–1997, and then plot this data on a scatterplot, as shown in Figure 4-22. We don't see any positive or negative drift in the scatter of points, which tells us that there appears to be no relationship between one's ability to hit opposite-handed pitchers (relative to same-handed pitchers) in 1998 and the corresponding opposite-handed hitting ability the previous four years.

The "Ability" Situations

Up to this point, we have shown that situational effects are essentially bias effects. All of the above situations, such as home/away and same-handed/opposite-handed appear to affect all players the same way. So for these situations, there appear to be few "situational stars"—players who take particular advantage of a given situation.

There are, however, two situations—the pitch count and the runners-on-base situation—that we haven't yet talked about, and as it turns out, these are probably the most interesting. For these situations, players appear to possess different true situational effects. Among the eight types of situational effects we are studying, it makes some sense only with these two to talk about unusually small

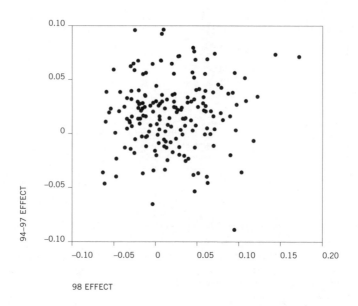

FIGURE 4-22   Scatterplot of 1998 situational effects and the previous four-year period situational effects for the same-handed/opposite-handed situation.

and unusually large individual player effects. When it comes to the pitch count and runners-on-base situations, it appears that individual ability varies significantly from player to player.

How did we decide that the pitch count situation, say, was different from the earlier six? We did try to fit a model which said that pitch count was a bias situation. That is, we tried to fit a model which said that each batter loses the same amount of hitting effectiveness when the pitch count is two strikes. But the model didn't fit very well in the sense that it did not predict the actual pitch count situational data that we see in *Player Profiles*. Instead, the observed pitch count situational effects we see in the book are more spread out than what one would expect if pitch count were really a bias effect. The same thing happened when we tried to fit a bias model to the scoring-position/none-on-out data. The actual data we see in the book have more variation than we would expect to see if this situation affected each hitter the same way.

Let's analyze the pitch count data first. Figure 4-23 shows a stemplot of the observed situational effects.

The center of these effects is at about 158 points. So players generally hit 158 points lower when the count is at two strikes (pitch counts 0–2, 1–2, 2–2, 3–2) instead of being ahead in the count. This is a *very* large effect. If you are watching a game and a pitcher gets two strikes on the batter, then it's pretty likely the batter will be heading back to the dugout in very short order.

What is even more interesting about this data than the size of the general effect is its range. Looking at the stemplot, some batters actually had a pitch count effect near zero—they hit for about the same average when they were ahead in the count or when they had two strikes. In contrast, two hitters had a pitch count of over 300 points! These hitters either were extremely good when they were ahead in the count or they were terrible when the pitch count got to two strikes.

Since the situational effects depend on ability, some players are better than others in handling different pitch counts. To find good and poor players in this situation, let's compare this data with the previous four years' data. For each player, we find the 1998 situational effect (AVG when ahead in count – AVG when two-strikes), and also find the same situational effect for the years 1994–1997. A scatterplot of the 1998 effects and the previous four-year period effects is shown in Figure 4-24.

What is notable about this graph is that there is a positive drift to the scatter of points. This means that there is a relationship between a player's 1998 effect

```
0 | 00111
0 | 233
0 | 4445555
0 | 666666667777777777
0 | 888888888899999999
1 | 00000000000000011111111
1 | 2222222222222223333333333333
1 | 4444444444455555555555555
1 | 66666666666666666677777777
1 | 888888999999999999
2 | 000000000000000000011111111111111111111111
2 | 2222233333333
2 | 444455555
2 | 6677
2 | 8888999
3 | 0
3 | 33
3 |
```

FIGURE 4-23   Stemplot of observed 1998 differences AVG(ahead in the count) – AVG(behind in the count).

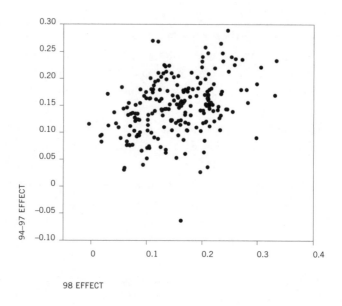

FIGURE 4-24 Scatterplot of 1998 situational effects and the previous four-year situational effects for the ahead-in-count/two-strikes situation.

and his 1994–1997 effect. Hitters who have high 1998 effects also tended to have high effects in the previous four-year period; similarly, low-effect hitters in 1998 tended to be low in the previous period. This confirms that the pitch-count effect is an ability-based effect, and not the result of chance. The relationship between the 1998 effect and the 1994–1997 effect isn't very strong—there is still a lot of scatter in the graph. But we didn't see this positive relationship for any other of the six situations that we previously analyzed.

Since the pattern of pitch count effects corresponds to real effects, it makes sense to pick out the players who are unusually high or low for both 1998 and the 1994–1997 periods. These players correspond to points which are in the upper right or lower left of the scatterplot. We've labeled some of the extreme points in the scatterplot in Figure 4-25.

There is a strong connection between a player's pitch-count situation effect and his likelihood of striking out. To see this, Table 4-11 lists six players in the lower left part of the plot who have small ahead-in-the-count/two-strike effects, and, at the other extreme, in Table 4-12, eleven players who have large pitch-count effects. The table gives the number of at-bats, number of strikeouts, and the strikeout rate for the players in the last five years. Players who have small

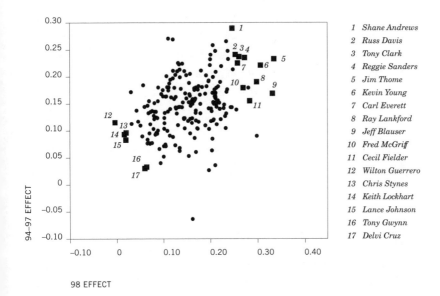

1	Shane Andrews
2	Russ Davis
3	Tony Clark
4	Reggie Sanders
5	Jim Thome
6	Kevin Young
7	Carl Everett
8	Ray Lankford
9	Jeff Blauser
10	Fred McGriff
11	Cecil Fielder
12	Wilton Guerrero
13	Chris Stynes
14	Keith Lockhart
15	Lance Johnson
16	Tony Gwynn
17	Delvi Cruz

FIGURE 4-25   Situational effects for the ahead-in-the-count/two-strike situation with scatterplot of 1998 situational effects and unusual players from the previous four-year period identified.

	At-Bats	Strikeouts	Percent
Shane Andrews	1151	344	30
Russ Davis	1201	314	26
Tony Clark	1659	429	26
Reggie Sanders	1964	552	28
Jim Thome	2214	625	28
Kevin Young	1360	335	25
Carl Everett	1442	339	24
Ray Lankford	2442	632	26
Fred McGriff	2697	521	19
Jeff Blauser	1956	419	21
Cecil Fielder	2287	563	25

	At-Bats	Strikeouts	Percent
Wilton Guerrero	761	117	15
Chris Stynes	672	57	8
Keith Lockhart	1263	125	10
Lance Johnson	2415	147	6
Tony Gwynn	2458	97	4
Delvi Cruz	890	110	12

TABLE 4-11   Players with Low Ahead-in-the-Count/Two-Strike Effects
TABLE 4-12   Players with High Ahead-in-the-Count/Two-Strike Effects

pitch count effects are relatively unlikely to strike out—their strikeout rates are in the 4–15 percent range. In contrast, the players who have high effects generally strike out about twice as often—these rates fall between 19 and 30 percent. Players in the two groups represent very different types of hitters. Tony Gwynn is representative of the first group—a hitter who has tremendous bat control and a short stroke. Jim Thome represents the second group of hitters—he has a long batting stroke that is well suited for power but at the cost of striking out a lot. So differences between these pitch-count effects are meaningful. They correspond to different batting styles, and the different abilities of players to control the bat.

How Large Are the True Ability Effects?

Another thing that we learn from the scatterplot of 1998 effects and 1994–1997 effects is that the true pitch-count effects are likely smaller than what we think based on the 1998 data. First, consider a player like Tony Gwynn. If we're interested in his true pitch-count effect, do you think it is better to use the 1998 data, or the data in the four-year period 1994–1997? Actually, it's better to use the four-year data, since it is based on a lot more at-bats. (Tony had 1997 at-bats in the 1994–1997 period compared to 461 at-bats in 1998.) In other words, Tony's pitch count effect in 1994–1997 is likely closer to his true effect than the effect that we observe in 1998.

We can learn about the relationship between the 1994–1997 effects and the 1998 effects by fitting a "best" line to the scatterplot that was graphed below. The equation of this best line is given by the following:

Previous Four-Year Effect = .1 + .3(*Effect in 1998*)

This equation tells you how to predict a player's previous four-year effect if you know his 1998 effect. Since the four-year effect is the best estimate of a player's true effect, this prediction is informative about the true effect.

Let's illustrate how this works. Suppose a player in 1998 bats 300 points better when he is ahead in the count as opposed to 2 strikes behind. That is, his batting average when he is ahead in the count is .3 larger. Using this equation, we predict that his four-year batting average advantage is:

$$.1 + .3 (.300) = .190$$

So his true advantage is actually more like 190 points. Suppose, on the other side, that a player in 1998 actually bats the same whether he is ahead in the count or two strikes behind. That is, his 1998 observed effect is 0. Using the equation, we predict his four-year batting average effect is:

$$.1 + .3\,(0) = .100$$

Although this person's pitch-count effect is below average, it is more likely to be 100 points than 0 points.

The moral of this discussion is that you shouldn't be deceived by large situational effects. Even when players appear to have different situational effects (such as in the pitch-count situation), the true effects are generally much smaller than the effects that we observe in a single season.

Game Situation Effects

Finally, let's talk about the last effect, which compares a player's batting average with runners in scoring position versus his batting average when the bases are empty and there are no outs. We'll call these the "clutch" effects, since they indicate how the players perform in a clutch, a stressful situation. First, let's look at the stemplot of the observed clutch effects for the 1998 season. (See Figure 4-26.)

The average effect is approximately 0. That is, about half of the players hit for a higher average when runners were in scoring position, and the other half hit better in the none-on/out situation. But we note a great spread in these effects—from −170 points to +220 points.

```
-1 | 7
-1 | 555444
-1 | 322
-1 | 1111000
-0 | 99998888
-0 | 777777776666666
-0 | 5555555555555444444444444444
-0 | 33333333333333322222222222222222
-0 | 1111111111111100000000000
 0 | 000000000000000000001111111111111
 0 | 222222222222222333333333333333333
 0 | 4444444444455555555555
 0 | 66666666666677777
 0 | 88899
 1 | 000011
 1 | 223333
 1 | 5
 1 | 66
 1 |
 2 |
 2 | 2
```

There is one simple explanation for this large range of effects. The players take only about half of their bats in one of these two situations. So these observed *scoring position* vs. *none on/out* effects are based on a relatively small number of at-bats. For this reason alone, there will naturally be more spread in these effects.

But this one explanation is not sufficient to account for all the variation we see in the stemplot. As it turns out, the variation is due to more than chance, and there is some evidence that players do possess different true clutch effects. We can check this by looking at the scatterplot in Figure 4-27, which plots the 1998 effects against the 1994–1997 effects.

FIGURE 4-26 Stemplot of observed 1998 differences AVG(Scoring Position) − AVG(None on/none out).

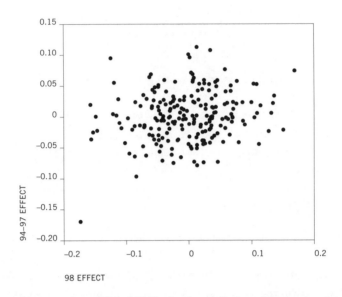

FIGURE 4-27   Scatterplot of 1998 situational effects and the previous four-year period situational effects for the scoring position/none on-out situation.

The points in this plot tend to drift slightly upward (as you move from left to right), which indicates that generally players with small clutch effects in 1998 tended to have small effects in the previous four-year period, and players who performed in the clutch in 1998 tended to perform in the clutch in 1994–1997. The value of the correlation between the 1998 and 1994–1997 effects is +.172, and is +.125 with the one unusual point in the lower left section of the plot removed. The relationship in this clutch-effect situation is weaker than the relationship that we saw earlier for the pitch-count situation.

Although there is evidence that players differ in their ability to perform in the clutch, the evidence is relatively weak, so we are reluctant to single out especially good and poor clutch hitters. Let's say at this point that a player might possess an ability to hit especially well or poorly in the clutch, but we don't quite have enough data to say who that player may be.

A Lot of Noise

When we open up *Player Profiles* and look at the batting averages broken down by different situations, we see a lot of fascinating high and low numbers. One player hits for a much higher average at home games, another player has a low average the first half of the season, a third player possesses a very poor average when he's behind in the count, and so on. These are intriguing discrepancies. They *must* be trying to tell us something.

Or maybe not. What we learned is that most of what we see in *Player Profiles* is, statistically speaking, mainly "noise," or random variation. That is, it's similar to what you get when you toss a coin. And when the situational batting average data you look at covers only a single year, it's very noisy. Only taken over a longer time span, such as the last five years, does it really start to tell us much.

Our basic model for batting averages is that players do indeed possess different true batting averages, and these true batting averages are bell-shaped about the typical value .276. That's where the normal curve peaks. When we think of the effect of a given situation, say home vs. away, our model suggests two bell-shaped curves for batting averages—one for the averages for games played at home, a second for games played away. Using this model, we placed situations into three categories.

First are the *no-effect situations,* the ones where no general pattern exists. Players seem to bat much the same before the All-Star Game as after. And they bat the same during day and night games. And the same on grass and turf fields. If you find a player that bats 80 points better during day games, we can probably find another who bats 80 points better at night.

Second are the *bias situations,* the ones where a general effect exists, but there is little evidence that individual players take advantage or disadvantage of a given situation. So, for example, we can say hitters generally bat 12 points better at home, and that appears true for most players. And managers are right to send up that right-handed pinch hitter against the lefty reliever. But beyond a few obvious and time-honored rules of thumb (including as well the higher batting average against ground-ball pitchers), there is not much to tell.

Finally, the pitch-count and clutch situational data are *ability situations,* since players seem to handle these situations differently. For example some players seem to bat the same no matter if the pitch count is 3–0 or 0–2, and other players bat much worse when the count goes to two strikes. There is some evidence that different players perform differently under pressure situations, although the 1998 data are not sufficient to identify players who are truly great hitters with runners in scoring position.

After all is said and done, Kenny Lofton does have a point: When he talks about the role of luck in hitting, his observations are pretty accurate. Over five years you will see trends and tendencies, but in a given situation or even in a complete single season, chance is what seems to rule.

STREAKINESS (or, The Hot Hand)

CHAPTER 5

One fascinating aspect of baseball is the widely-held belief, among fans as well as the media, that players can be "streaky." It's often said of a particular player that in the last few games, or weeks, or even months, he's had a "hot" or "cold" hand. And it seems entire teams can go on a streak, as we found when we did an Internet search for the words "streaky" and "hot hand":

> "Rays Streaky in Spring"
> —*Florida Sports Network*, March 24, 1998

This article describes the Devil Rays' tendency to play a little streaky—citing two losing streaks of six and four games in their 1998 spring exhibition season. The writer cautions the fan not to read too much into these losing results, and the manager is quoted as saying that at this point in the season, he is more concerned about individual performances than team results.

> "Braves Have No Trouble Beating Streaky Pirates"
> —*Observer-Reporter* (Washington, Pa.), May 19, 1999

In this article, the hitting heroics of a "slumping" Chipper Jones are described—he went on a 9 for 50 (.180) slide before this game. The article also noted that the Braves' Bret Boone reached base by a hit or a walk in 20 straight games. The reference to the "streaky Pirates" in the title refers, of course, to the fact that they have been playing unusually well in the games before the present one.

> "Streaky Cal Softball Team Gets Two Splits"
> —*The Daily Californian* (Berkeley, Cal.), May 4, 1998

The writer of this piece seems to be saying that there were two completely different Cal softball teams on the field in two doubleheaders one weekend. In the two victories, Cal's hitters were described as "phenomenal," and in the two losses, Cal was unable to get any solid contact with the bat. The Cal catcher describes their hitting as contagious: "If someone gets a big hit, the rest of us go out thinking, 'If she can hit it out, so can I.' It's such a mental game." The writer also says that the Bears have been plagued by inconsistencies at the plate all year long.

In an ESPN.com profile of Todd Zeile, the Texas Ranger third baseman is described as a streaky hitter. He typically starts slowly and does his best hitting late in the season, says the analyst for the website.

Finally, In Mike Zaidlin's "Thinking Baseball" column on the World Wide Web (www.thinkingbaseball.com), the author criticizes Don Zimmer's strategy of playing hitters who appear to have a "hot hand." Zaidlin thinks that Yankees management puts too much faith in the notion of a "hot hand," because it is based on too small a number of at-bats.

Thinking about Streakiness

What are these writers talking about when they say a team is on a hot or cold streak, or a player has a hot or a cold hand? One thing they obviously mean is that the team or player is going through an unusually long stretch of good (winning) or bad (losing) behavior. If a hitter like Chipper Jones bats only 9 for 50, we say that he has a cold hand because he normally hits for a much higher average, and 50 at-bats appears to be a long time for him to go with only 9 hits. And if a team like the Devil Rays has several long runs of uninterrupted losses, as well as long runs of uninterrupted wins, we say the team is streaky, since it seems to be winning or losing "clusters" of games.

Most of the articles above seemed to be talking about the streaky and hot-hand *performances* that were observed during a season. But there is a second, deeper meaning of streakiness: sometimes the word is used to describe the *nature* or *ability* of a player or team. A player such as Todd Zeile may be called streaky since people believe that his true batting ability is streaky. On some days, says this theory, Zeile feels very comfortable with the bat, and he has a high chance of getting lots of hits. Then, on other days, Zeile's batting stroke seems to be out of sync, and he has a much smaller chance of getting a hit. Similarly, we might say that a team is on a hot streak when we believe that as a group the players are performing to the best of their ability and the team has a high probability of winning. (We could also describe this situation by saying that the team is "in a groove" or even "on a tear.") At other times, the team may have problems, such as injuries or dissension in the clubhouse, that we believe have an adverse effect on performance, and so the team has a small probability of winning. (Then the team is "slumping" or has "gone cold.")

Fans and sportswriters frequently confuse or don't distinguish between the *performance* and *ability* interpretations of streakiness. When a team goes through good spells and bad spells, clusters of winning and clusters of losing, we are observing streaky behavior. It is natural in such situations for the fan or sportswriter to provide some rationale for the observed streakiness: the performance is explained by something in the nature of the team. For example, one might say that a team's tendency to go on cold streaks is due to the inexperience of the players, the inconsistency of the starting pitchers, the tactics of the manager, and so on. But, as we will shortly see, it's possible that the team *is not* really streaky by nature, but due to chance or luck, they appear to be performing streakily.

One goal of this chapter is to clearly distinguish between a player or team's true streaky or hot hand *ability* and the streaky or hot hand *performance* that we observe during a baseball season. We first will discuss some common mistakes that people make in interpreting baseball averages. Then we will focus on the first-half batting performance of Todd Zeile in the 1999 season. Looking at his hitting record, we will notice several interesting patterns that indicate that Zeile may be a streaky hitter. Next, we will propose several models for Zeile's hitting ability. One model, which we will call Mr. Consistent, says that Zeile is the

Batter	Pos	AB	R	H	RBI	AVG
Knoblauch	2B	3	0	1	0	0.333
Jeter	SS	3	2	3	1	1
O'Neill	RF	3	0	0	0	0
Williams	CF	4	0	1	1	0.25
Martinez	1B	4	0	0	0	0
Davis	DH	2	1	1	1	0.5
Ledee	LF	4	0	0	0	0
Brosius	3B	3	0	0	0	0
Giradi	C	3	0	0	0	0

TABLE 5-1   Box Score of the New York Yankees in Their First Game of the 1999 Season

ultimate consistent hitter—he comes to every at-bat with the same chance of getting a hit. Then we'll work with a very different model, called Mr. Streaky. In this model, Zeile is either hot or cold during a single at-bat where the chance of getting a hit when he is hot is a large number and the chance of hitting when he is cold is small. In addition, in the Mr. Streaky model, if Zeile is hot on a particular at-bat, he is very likely to remain hot in the next at-bat.

Once we have described ways of measuring streaky performance and models for Zeile's true hitting behavior, we show how we can learn about Zeile's hitting behavior (the model) on the basis of his hitting data (the statistics). We extend our basic method to the win/loss records of the 30 major league teams in 1998. Some of these teams performed very streakily in 1998, and we will see which teams actually seem to possess true streakiness based on their season performances.

Interpreting Baseball Data

Before we talk in more detail about streakiness, it will be helpful to describe some basic difficulties that people have in interpreting baseball statistics. We focus on interpreting a batting average, although the difficulties we describe will apply to interpreting any baseball statistic.

Let's suppose that your favorite player is Tino Martinez of the Yankees. The Yankees opened the 1999 season at Oakland, and, to learn about how your favorite player performed in his first game, you look at the published box score, shown in Table 5-1.

We see that Tino went 0 for 4 in the game, so his current 1999 batting average (based on this single game) is 0/4 = .000. Now the typical fan is interested in drawing some conclusion about Tino's batting ability on the basis of this .000 average. Can the fan conclude that Tino's in a slump? That is, can the fan conclude that Tino is a slow starter and his swing is a little rusty?

Here the fan is interested in making a statement about Tino's batting ability from this game's hitting statistics. We can measure Tino's batting ability in terms of a probability p. This number is the chance that Tino gets a base hit on a single at-bat. The fan is interested in saying something about Tino's hitting probability based on his 0-for-4 game performance. Of course we don't know Tino's hitting probability, but we can make an educated guess at this probability based on his hitting record in his previous nine years in the major leagues (see Table 5-2).

We see that Tino's seasonal batting averages generally increased over time, hit a peak in 1997, and dipped slightly in 1998. If we make the assumption that

Year	AB	H	AVG
1990	68	15	.221
1991	112	23	.205
1992	460	118	.257
1993	408	108	.265
1994	329	86	.261
1995	519	152	.293
1996	595	174	.292
1997	594	176	.296
1998	531	149	.281

TABLE 5-2   Hitting Statistics for Tino Martinez in the First Nine Years of His Major League Career

Tino is a little bit past his prime as the 1999 season begins, it is reasonable to assume that his hitting probability in the opener against Oakland is $p = .280$.

If one thinks of Tino as a true .280 hitter, many baseball fans will have trouble predicting how Tino will hit during the season. To start off, how many hits will Tino get in a game where he has four at-bats? Well, many Tino fans will think that their man will go 1 for 4 in this game, since .280 is close to .250 = 1/4. Moreover, these fans will be unpleasantly surprised if Tino goes hitless (as in the game described above), or pleasantly surprised if he goes 2 for 4.

Actually, although it is *most likely* that Tino will go 1 for 4 in this game, the probability that he gets exactly one hit is only 42 percent. So actually, it is more likely (58 percent) that Tino will *not* get one hit. Moreover, there is a sizeable probability (27 percent) that Tino will go hitless. So the above game result (0 for 4) is entirely consistent with Tino being a .280 hitter, and there is no reason to think that he is in a slump.

Now, maybe you are not surprised by the above comments. After all, you can't learn much about a player's true batting average on the basis of one game. But suppose you watch Tino's hitting for the first seven games of the season, as shown in Table 5-3. For these seven games, Tino was 4 for 25, for a .160 average. Now, for most fans, Tino appears to be a slump. Intuition tells us he is clearly *not* a .280 batter for this first week in the season.

But this intuition is wrong. There is a reasonable chance that a true .280 hitter will have a slump like this one. If a true .280 hitter comes to bat 25 times, as Tino did in his first seven games, the chance that he will get 4 or fewer hits is 13

Date	Apr 5	Apr 6	Apr 7	Apr 9	Apr 10	Apr 11	Apr 13
H/AB	0/4	1/5	1/4	1/1	0/3	0/4	1/4

TABLE 5-3   Batting Data for Tino Martinez for the First Seven Games of the 1999 Season

percent. Now 13 percent is not a high probability, but if you watch seven .280 hitters bat for one week, there is a high probability that you'll see one of them get 4 or fewer hits. So we shouldn't be too surprised by Tino's 4-for-25 stretch. He may really be a .280 hitter but, by chance variability, he was unlucky during this first week of the season.

The moral here is that one has to be very cautious in interpreting baseball averages from a small number of at-bats. It is common for fans to believe in the so-called law of small numbers. This law says that one will observe what one expects, even over a small sample size. So if a player like Tony Gwynn is a .333 hitter, you think that he will get 1 out of 3 hits in every game. The law of small numbers isn't true. Even if Gwynn is a true .333 hitter, it is very likely you will observe him hit for significantly lower and higher averages if your observations are based on a small number of games.

Moving Averages—Looking at Short Intervals

It's clear from the preceding discussion of Tino Martinez's batting average that one needs to be careful about drawing conclusions from very limited sets of data. The problem is, when you are talking about streaks, you are often talking about relatively brief bursts of activity (or clusters of inactivity)—an eight-game hitting streak by a .220 hitter, or a five-game stretch where a .300 hitter can't get on base. To get a sense of how we might deal with the statistics of streakiness, we'll take a look at Todd Zeile's hitting statistics for the first half of the 1999 season. (We are looking at only the first half of his season because that's when this book was written.) Table 5-4 gives the game date and the number of hits and at-bats for each of the first 80 games that Todd played. Two particular streaks— one hot and one cold— are in boldface.

For the 80 games shown, Todd had 84 hits in 301 at-bats, for a batting average of 84/301 = .279. Now, if we look at this table carefully, we'll see short time periods where Todd was unusually hot and cold. For the eight-game period from April 15 through April 23, Todd had a tough hitting stretch where he only got 2 hits in 29 at-bats:

Date	Hits/AB							
Apr 5	1/4	Apr 29	1/4	May 22	1/4	Jun 14	1/3
Apr 6	2/3	Apr 30	1/4	May 23	0/3	Jun 16	1/2
Apr 7	3/4	May 1	1/4	May 24	2/5	Jun 17	1/3
Apr 9	1/4	May 2	0/2	May 25	2/4	Jun 18	0/3
Apr 10	0/2	May 3	0/4	May 26	2/4	Jun 19	1/4
Apr 11	2/4	May 5	1/4	May 28	1/4	Jun 20	0/4
Apr 12	2/4	May 6	2/4	May 29	3/4	Jun 21	1/4
Apr 13	0/4	May 7	0/4	May 30	0/2	Jun 22	1/4
Apr 14	1/3	May 8	0/4	May 31	1/4	Jun 23	0/4
Apr 15	**0/4**	May 9	3/4	Jun 1	1/3	Jun 24	1/4
Apr 16	**0/2**	May 10	0/4	Jun 2	0/4	Jun 25	1/6
Apr 17	**0/3**	May 11	1/4	Jun 4	0/4	Jun 26	2/4
Apr 18	**1/5**	May 13	2/5	Jun 5	0/4	**Jun 27**	**3/4**
Apr 20	**0/3**	May 14	0/4	Jun 6	0/4	**Jun 28**	**1/3**
Apr 21	**1/4**	May 15	2/4	Jun 7	3/3	**Jun 29**	**3/5**
Apr 22	**0/4**	May 16	1/3	Jun 8	1/5	**Jun 30**	**4/4**
Apr 23	**0/4**	May 17	0/3	Jun 9	0/3	**Jul 2**	**1/4**
Apr 25	2/5	May 18	1/3	Jun 11	0/4	**Jul 3**	**2/5**
Apr 27	1/3	May 19	1/4	Jun 12	2/5	**Jul 4**	**1/3**
Apr 28	2/4	May 21	1/3	Jun 13	1/4	**Jul 5**	**2/3**

TABLE 5-4   Todd Zeile's Batting Statistics for the First Half of the 1999 Season

$$\frac{2}{29} = .069$$

In contrast, look at the period from June 27 through July 5, when Todd was on fire, getting 17 hits in 31 at-bats:

$$\frac{17}{31} = .548$$

So, although one could reasonably call Zeile a .279 hitter, he hit for much smaller (.069) and much larger (.548) averages over short time periods.

A *moving average* plot, shown in Figure 5-1, is an effective way of displaying these short-term batting averages. This graph plots the short-term batting average of Zeile over all groups of eight adjacent games. (In the language of statistics, this is called a moving average with a width of 8.) In games 1 through 8,

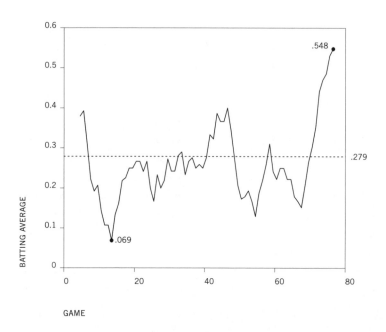

FIGURE 5-1   Moving average plot of Zeile's hitting data using a width of eight games.

Zeile got 11 hits in 29 at-bats for a .379 average. In Figure 5-1, this average (.379) is plotted against the mean game number (4.5 is the mean of 1, 2, . . . , 8). Moving one game ahead, we look next at the group of eight games numbered 2 through 9. There, Zeile had 11 hits in 28 at-bats, and the batting average (.393) is plotted against the mean game number (5.5). We continue in this way, stepping ahead one game at a time, until we get to the final group of eight games, numbered 73 though 80.

In Figure 5-1, we use dots to call out the two particularly interesting eight-game stretches noted earlier. Also, we show Zeile's season batting average (.279) as a dashed horizontal line.

This graph is a good way of displaying Zeile's pattern of hitting for the first half of the 1999 season. We see that, after an initial hot streak, Todd had a slump in the first part of the season, then a good hitting stretch from games 40–50, two minor slumps from games 50–70, finishing with a hot streak at the end of the period. Generally, it *seems* remarkable how much spread there is in these eight-game batting averages. But to get a better sense, we can measure how much streakiness we see in this plot by calculating the difference between the largest and smallest moving average:

Largest 8-game average – smallest 8-game average

Here this difference is .548 – .069 = .479, which appears large. Zeile hit 479 points better on his best 8 games than his worst 8 games.

Runs of Good and Bad Games

Another way to describe Zeile's hitting behavior is to look for interesting patterns of good and bad games across time. Remember, Zeile's batting average for all 80 games is .279. We will say that Zeile's batting for a particular game is good is if his average on that day exceeds .279; otherwise we'll say he had a bad hitting game. So, if Todd bats 2 for 4 (.500) for a game, we'll call it a good game and denote it by a "+"; and if he hits 1 for 4 (.250), we'll call it bad and denote it by a "0". In Figure 5-2, we have classified the 80 games.

One interesting pattern in the sequence of good and bad days is a *run,* a streak of consecutive bad games (like "00000") or good games ("++++"). The first part of the Zeile game-hitting sequence is "0++00++". Looking at Figure 5-3, we see that Todd started with a run of one bad game, a run of two good games, a run of two bad games, and a run of two good games. Two interesting runs are underlined in the figure. In each case, Zeile had eight consecutive bad hitting games. It's also interesting that Zeile followed his second eight-day hitting slump by hitting well in eight out of nine games.

So one interesting pattern is a *long run* of good or bad hitting games. These long runs indicate that Zeile might be a streaky hitter. Another thing that we can compute is the *total number of runs of good and bad games*. Let's suppose that Zeile is really a streaky hitter. Then we would expect him to follow good hitting days with good days and likewise bad hitting days with bad days. ("When

	April	*May*	*June*	*July*
DATE	1111111111222222223	1111111112222222233	1111111122222222223	
	5679012345678012357890	1235678901345678912345678901	1245678912346789012345678	02345
HIT	0++00++0+00000000+++00	0000+00+00+0++0+0+00+++0+00	+0000+000+0+++00000000++++	+0+++

FIGURE 5-2   Classification of Zeile's games into good (+) and bad (0) hitting games.

0++00++0+<u>00000000</u>+++000000+00+00+0++0+0+00+++0+00+0000+000+0+++<u>00000000</u>+++++0+++

FIGURE 5-3   Identification of two long runs in Zeile's hitting sequence.

```
0++00++0+00000000+++000000+00+00+0++0+0+00+++0+00+0000+000+0+++00000000+++++0+++
12  3 4  567      8   9       11  11  111 11122 2    222 22  22   333  3       3       33
                  01  23  456 78901 2    345 67  89   012  3      4       56
```

FIGURE 5-4   Counting the number of runs in Zeile's hitting sequence.

Number of Game Hits	0	1	2	3	4
Count	26	32	15	6	1

TABLE 5-5   Count of the Number of Games in Which Zeile Had 0, 1, 2, 3, and 4 Hits

you're hot, you're hot, and when you're cold, you're cold.") In this case, there will generally be many runs of long length and few runs of short length, and the total number of runs in the sequence will be small.

We've counted the number of runs in the Zeile sequence in Figure 5-4 by labeling the beginning of each run with a number. We see that there are 36 runs in this hitting sequence. Is 36 a small number of runs? Actually, at this point of the discussion, we don't know, but we'll come back to this question later.

Numbers of Good and Poor Hitting Days

We can also look at Zeile's hitting data by counting his hits per game, then categorizing each of the 80 games according to hit count. Table 5-5 shows the number of 0-hit games, the number of 1-hit games, etc. We see that Todd had seven games in which he had either 3 or 4 hits, and 26 games (out of 80) in which he was hitless. If these numbers seem high to the average fan, they provide evidence that Zeile was a streaky hitter.

What Is Zeile's True Hitting Ability?

So in our look at Zeile's hitting data for the 80 games, we saw some interesting features. Todd had some unusually small and large batting averages over short time intervals, he had several long runs of bad hitting days, and he had some games in which he hit especially well. At this point, the question is: Should we be surprised by these observations? Do these data suggest that Zeile was really a streaky hitter during the first half of 1999?

We will try to answer this question by proposing some simple models for Zeile's true batting ability and see what we learn about these simple models on the basis of Zeile's hitting record in 1999. For simplicity, we assume only two distinct models, although the basic method can be used to distinguish between a large number of models. We will describe this statistical method in three parts.

First, we'll assume that Zeile is really a consistent player who comes to every at-bat with the same chance of getting a hit. Also, we will assume that his chance of hitting on a given at-bat is not influenced by what he did on previous at-bats. This type of player is the ultimate "Mr. Cool." If Zeile is this type of player, we'll look at the kind of batting record he will achieve in an 80-game schedule.

Next, we'll assume that Zeile really is a streaky hitter. We carefully define what we mean by streaky. We'll assume that the chance that he gets a hit may change on different at-bats. We will say that, in a particular game, Zeile is either "hot" with a large chance of getting a hit, or "cold" with a much smaller chance of hitting. Moreover, if Zeile is hot in a given game, he is more likely to remain hot (than become cold) for the next game. Likewise, cold at-bats are more likely to be followed by cold at-bats. If Zeile is this type of streaky hitter, we will see how he would perform in an 80-game season.

Finally, we will talk about how we learn about Zeile's true hitting ability (consistent or streaky) based on the patterns we saw above. Suppose, for example, that we observe two long runs of bad hitting games. Based on these data, what is the chance that Zeile is a consistent hitter and what is the chance that he is a streaky hitter?

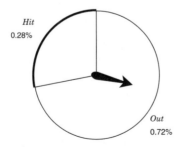

FIGURE 5-5   Spinner for a consistent hitter.

Mr. Consistent

Most people believe that Todd Zeile is a streaky hitter. But what if Zeile were not streaky? In fact, suppose that Zeile was the ultimate non-streaky, or consistent, hitter? What would that mean?

We use the basic spinner model to represent Zeile's success (or failure) in a single at-bat. (Here we are just considering official at-bats, ignoring events such as walks and sacrifices.) We visualize a spinner with two regions designated "Hit" and "Out. " (See Figure 5-5.) The Hit region corresponds to the chance that Zeile will get a hit on the single at-bat.

Suppose that the *same spinner* is used for every at-bat during the season. This means that Zeile has the same chance of getting a hit over all 80 games. This assumption probably seems far-fetched. You might think that the chance of getting a hit will depend on a number of factors, including the pitcher, the ball park, the game situation, and so on. You may be right. But let's pretend for now that the hitting probability doesn't depend on these factors. We think of Zeile as a hitting machine who actually has the same chance of getting a hit every time he comes to bat. We'll call this hitting probability p.

Another important assumption made here is the *independence* of hitting results of different at-bats. Suppose that the spinner is spun and Zeile gets a hit on his first at-bat. This successful result will have no bearing on what happens on the next at-bat. One aspect of our spinner is that it has no memory—it doesn't remember how many times a Hit or Out was spun in the past. So the chance of getting a hit on a particular at-bat will be the same number p no matter if Zeile has done well or poorly in his previous at-bats. This property of *independence* is actually the opposite of streakiness, and instead represents the hitting characteristics of a player that we'll call "Mr. Consistent." This player is the ultimate "cool customer," who has the same chance of getting a hit under all possible circumstances.

How Does Mr. Consistent Perform During a Season?

Suppose that Zeile really was a consistent hitter. What would his hitting data for the 80 games look like? We can answer this question by simulating from the spinner model. We assume that Zeile's chance of getting a hit on every single plate appearance is $p = .280$ (close to his .279 average for the 80 games in 1999). Then we simulate the results of all 80 games using the actual at-bat numbers of Zeile for the first half of the 1999 season. (We actually do this simulation on the computer, but it is equivalent to using our random spinner many times.)

Figure 5-6 displays moving average plots for Zeile's data and hitting data for eight other simulated players using our Mr. Consistent model. Of course, the center plot in the figure should look familiar—it's Zeile's moving average plot that we saw earlier. How do the consistent hitter graphs compare with the graph of Zeile? Actually, what stands out is the large spread (up and down pattern) of the moving averages of the consistent hitters. An extreme situation is the hitting of the consistent hitter at the lower right. Remember, this guy is a true consistent hitter; he gets a hit with probability .280 on every single at-bat. Nonetheless, he appears very streaky during this 80-game stretch. He is very hot in the first part of the season, gets cold in the middle, then is somewhat hot, and seems to fade near the end of the half-season. The player really is consistent—our use of the spinner model guarantees that—but his hitting performance during these 80 games is very streaky. The hitting patterns of the other seven players aren't quite so volatile as the one in the lower right, but all of the

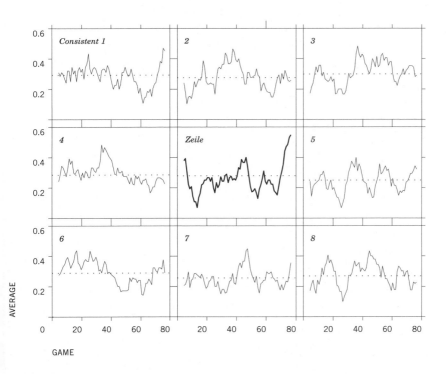

FIGURE 5-6   Moving average plot of hitting by Zeile and eight simulated consistent hitters with a hitting probability of $p = .280$. In all graphs, moving averages are computed using adjacent groups of eight games.

hitters show some streaky behavior. In other words, consistent hitters *can appear* very streaky in their hitting performances.

We can describe the streakiness that we see in these graphs using the same statistics we used to describe Zeile's data. These statistics include:

- The difference between the largest and smallest moving averages, using groups of eight games. (We will call this statistic MAX – MIN.)
- The number of long streaks (either good or bad), where "long streak" is defined as eight or more consecutive games.
- The number of runs in the sequence of good and bad games.
- The number of hitless games.
- The number of games with three or more hits.

Table 5-6 gives the values of these five statistics for each of the simulated seasons. To see if Zeile is different from a truly consistent hitter, we see if his statistics appear different from the statistics of the others. To make this comparison easier, we have computed the mean (arithmetic average) of each type of statistic for the consistent hitters

Let's illustrate this comparison using the statistic "Maximum moving average – Minimum moving average" shown in the first column of the table. Recall

	Max – Min	Number of long streaks (8 or longer)	Number of runs	Number of 0-hit games	Number of 3+-hit games
Zeile	0.479	2	36	26	7
Consistent 1	0.362	1	42	23	7
Consistent 2	0.361	1	41	26	7
Consistent 3	0.316	0	41	21	5
Consistent 4	0.317	0	43	21	4
Consistent 5	0.333	0	42	28	4
Consistent 6	0.295	0	34	20	6
Consistent 7	0.297	1	38	26	3
Consistent 8	0.333	1	42	25	5
Mean	0.327	0.5	40.4	23.8	5.1
p-value	0.010	0.06	0.26	0.34	0.23

TABLE 5-6   Statistical Values for Todd Zeile and Eight Simulated Consistent Hitters

that Zeile had a maximum moving average (using eight-game groups) of .548 and a minimum moving average of .069 for a difference as follows:

$$\text{MAX} - \text{MIN} = .548 - .069 = .479$$

Should the value .479 surprise us? Well, in order to make a judgment on that we have to look at the MAX – MIN values for our eight simulated consistent hitters:

$$.362, .361, .316, .317, .333, .295, .297, .333$$

The simulated players' MAX – MIN values range between .295 and .362, with a mean value of .327. Zeile's value, .479, is larger than all of them. So it appears that Zeile's difference between his best and worst moving average is larger than one would anticipate if Zeile really was a consistent hitter. Thus, there is some evidence that Zeile is not a consistent hitter, although we really can't say at this point that he is streaky.

In a similar fashion, we can look at the other four statistics to see if Zeile's value is similar to the values for the consistent hitters.

- Zeile had two long streaks (of eight games or more). This statistic appears unusually large, since none of the simulated consistent hitters had more than one long streak.

- We observed 36 runs in Zeile's sequence — is the number 36 unusually small for a consistent hitter? We would say yes since only one out of the eight consistent hitters had 36 or fewer runs.

- We thought Zeile's 26 hitless games statistic was large, but three of the eight simulated players had 26 or more hitless games. So the large number of hitless games doesn't appear unusual for a consistent hitter.

- Finally, we thought Zeile's seven games of 3+ hits were large, but two of the eight simulated hitters had seven or more games with 3+ hits. As with hitless games, this statistic does not definitively set Zeile apart as a streaky player.

We can't draw very strong conclusions from Table 5-6 since we only did the simulation eight times. We would do better if we simulated hitting data from the consistent model a large number of times, and then made a call on whether Zeile's hitting behavior fit into the Mr. Consistent model.

To check this out, we simulated data from a large number of consistent hitters (1000), and for each simulated hitting season we recorded values of the five statistics shown in Table 5-6. To see if Zeile's hitting statistics conforms to this model, we compute a p-value. This is the probability, assuming a consistent model, of observing a value of the statistic at least as extreme as the Zeile value. To illustrate, we observed a MAX − MIN value of .479 for Zeile. Using the simulated hitting data, we find the following:

p-value = Pr(MAX − MIN value is at least as large as .479) = .01

Thus, if we assume this consistent model, the chance of observing a MAX − MIN moving average difference of .479 or greater is only .01, or 1 percent. Thus it is safe to say that Zeile's data appear different from hitting data generated from a consistent model.

Likewise, we look at the p-value row of Table 5-6 to check for agreement of Zeile and the simulated consistent hitters with respect to the other four statistics. The only other p-value that appears unusually small corresponds to "long runs." Zeile's two long runs are unusual for hitters who are truly consistent. The p-values for the other three statistics are in the .23–.34 range. These statistics (number of runs, number of hitless games, and number of 3+-hit games) for Zeile seem to agree with the statistics for the simulated consistent hitters.

So, what have we learned? Even if a hitter is truly consistent, with the same chance of getting a hit on every single at-bat, his batting performance across 80 games can look pretty streaky. Even so, Zeile's hitting performance looks a bit different from the performance of true consistent hitters. The statistics that seem to stand out for Zeile are the great range between good and poor short-run batting averages (MAX − MIN) and the two long runs of bad hitting games.

Mr. Streaky

In the above discussion, we gained some understanding of Zeile's hitting performance by assuming he really was consistent, then seeing how he did and did not seem to fit into the Mr. Consistent model. But what if Zeile really was a streaky hitter? What does it mean to be streaky? And how do streaky players perform during a baseball season?

First, to be streaky, a hitter must have at least two possible hitting states. For simplicity, we'll assume that there are exactly two, which we'll call "hot" and "cold." When a hitter is hot, his hitting mechanics are great, he sees the ball well, and he has a high probability of getting a base hit. We will denote this probability as p_H. In contrast, a "cold" hitter is struggling with his hitting motion and is not swinging well. In this cold state, the batter has a small probability p_C of obtaining a hit.

As mentioned earlier, Todd Zeile's true batting average for 1999 is around .280. If Zeile really is a streaky hitter, there will be a big difference between the chance of getting a hit when he's hot and the chance when he's cold. We will assume that Zeile hits 100 batting points better than average when he's hot, and 100 points lower than average when he is cold. Accordingly, we will assume that $p_H = .380$ and $p_C = .180$. The hot hitting probability is similar to the average of Tony Gwynn in his best hitting season; the cold probability is similar to that of a weak-hitting shortstop who's in the Major Leagues because of his defensive ability. (You might not agree with the numbers we've assigned to a hot hitting probability and a cold hitting probability, but there should be a significant difference between the two.)

So one basic assumption about our truly streaky hitter is that in some games he hits for a high probability and in others he hits with a small probability. A second assumption describes how the streaky hitter moves between hitting states for different games. A streaky hitter has the tendency to stay hot for a number of games. If a streaky player is hot for one game, then it is likely that he'll remain hot for the next game (and unlikely that he'll change to cold). In other words, "If you're hot, you're hot!" Likewise, if a player is a cold hitter one game, then he will likely stay in a cold state in the next game. We will let the letter s (for "stay") denote the probability of staying in the same state from one game to the next:

$$p(\text{hot in second game if hot in first}) = s$$
$$p(\text{cold in second game if cold in first}) = s$$

To be streaky, it makes sense to let the probability s be a large value, like $s = .9$, which means that the hitter is likely to remain in the same state. Figure 5-7 illustrates the probabilities of shifting between hot and cold states for successive games. Note that if the chance of staying in the same state is .9, the chance of switching states (from cold to hot or hot to cold) is .1.

If a player is streaky in the manner we just described, how will he hit during a season? We learn about his season hitting by means of a simulation like the one done for the consistent hitter. This simulation is a little more complicated to run, however, since the probability of getting a hit can change from game to game.

Here's how it works. First, we visualize two spinners (shown in Figure 5-8), one to use when the hitter is hot, and the second to use when the hitter is cold. The spinners differ with respect to the sizes of their Hit and Out regions. For the Hot spinner, the hitting area is p_H (.380); for the Cold spinner it is p_C (.180).

Suppose we want to simulate Mr. Streaky's hitting for his first ten games: the results of this simulation are shown in Table 5-7. To start off, we flip a coin to

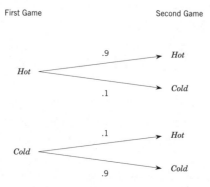

First Game   Second Game

FIGURE 5-7   Probabilities of changing states for a streaky hitter.

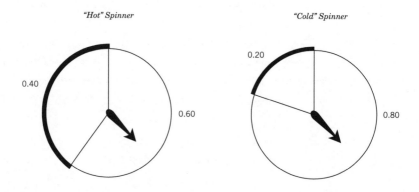

FIGURE 5-8   Spinners for a streaky hitter.

decide the player's hitting state for the first game; if the coin is heads, he'll be hot, and if the coin lands tails, he'll be cold. We observe heads, which means that Mr. Streaky is a hot hitter for this first game. We grab our Hot spinner and spin it for his 4 at-bats in his first game. The spinners lands in the Hit region twice, which means that he was 2 for 4 in his first game.

Let's move on to the second game. We use the switching probabilities to determine the state of Mr. Streaky for this game. He was hot in game 1, so he'll remain hot in game 2 with probability .9 and switch to cold with probability .1. We use a random spinner, as pictured in Figure 5-9, to determine if Mr. Streaky stays in his current hot state or switches to cold. In this particular simulation, the result of the spinner is Stay, so Mr. Streaky will be hot in game 2.

Now that we know that the hitter is hot in game 2, we use the Hot spinner to simulate hitting. Table 5-7 indicates that he has a tough game 2, going hitless in

0.10
Switch

0.90
Stay

FIGURE 5-9   Spinner to decide on switching or staying in current hot or cold state.

Game	1	2	3	4	5	6	7	8	9	10
State	Hot	Hot	Hot	Hot	Hot	Hot	Hot	Cold	Cold	Cold
Hitting Probability	0.38	0.38	0.38	0.38	0.38	0.38	0.38	0.18	0.18	0.18
AB	4	3	4	4	2	4	4	4	3	4
H	2	0	1	3	1	1	1	0	1	0

TABLE 5-7   Simulation of Mr. Streaky's Hitting for Ten Games

three at-bats. We continue in this manner to simulate the results of the remaining games. We use the switching/staying spinner to decide the state of a game, and then the Hot or Cold spinner to determine the hitting results of that game. Refer again to Table 5-7 to get the simulated hitting results for all ten games.

The results of this particular simulation are interesting. There is a clear pattern in Mr. Streaky's hitting states—he was really hot (hitting with a high probability) in the first seven games and cold (hitting with a small probability) for the remaining three. However, it is difficult to detect this true hitting behavior by just looking at his hitting statistics. For example, Mr. Streaky was 0 for 3 and 1 for 4 (twice) on days where he was a true hot hitter.

How Does Mr. Streaky Perform During a Season?

Earlier we looked at how truly consistent hitters would perform during an 80-game season. How would truly streaky hitters perform in the same span of games? First, we use the streaky model described above to simulate data for eight hitters. Figure 5-10 shows moving average plots of the batting averages of Todd Zeile (middle graph) and our eight simulated streaky hitters. If we compare these moving averages with those of the consistent hitters of Figure 5-6, we generally note more up and down behavior in the streaky graphs. That is, it seems that hitters who are truly streaky, in the way we have defined them, will tend to have unusually high and low batting averages over short stretches of games. But, this is not always the case—for example, the moving average plot of the streaky hitter in the upper center graph of Figure 5-10 looks pretty flat. Although this player is really streaky, he had a pretty consistent hitting pattern over the season.

Next, to measure the streakiness that we observe in these graphs, we compute the same five statistics that we used earlier to describe the streakiness we saw in Zeile's hitting. The values of these statistics for our eight simulated streaky hitters are given in Table 5-8.

First, let's compare the statistics for the eight consistent hitters (Table 5-6) with the corresponding statistics of the streaky hitters in Table 5-8. The streaky

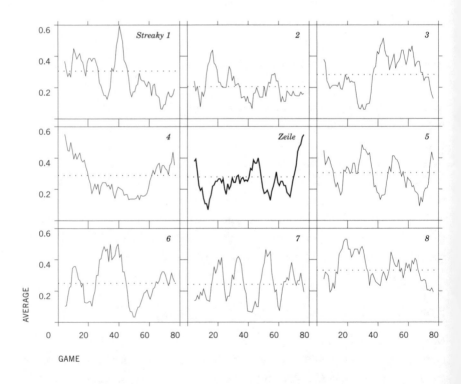

FIGURE 5-10   Moving average plot of hitting by Zeile and eight simulated streaky hitters with hot and cold hitting probabilities of $p_C = .180$ and $p_H = .380$. In all graphs, moving averages are computed using adjacent groups of eight games.

hitters tend to have larger values of Max − Min (the difference between the largest and smallest moving averages), a smaller number of runs, a larger number of hitless games, and a larger number of 3+-hit games. It is harder to say that the streaky hitters tend to have a different number of long streaks based on this small amount of simulated data.

Does Todd Zeile's hitting data agree with the hitting data for our simulated streaky hitters? To make this comparison, we simulate 1000 80-game seasons using our streaky model. For each season, we compute the five streaky statistics; the p-value row of Table 5-8 gives (for each statistic) the probability that a truly streaky hitter obtains a statistic value as least as extreme as Zeile's value. In Table 5-9, we compare these p-values with the ones that we obtained earlier for the simulated consistent hitters.

Note that for each one of the five game statistics, the p-values are larger for the streaky model. This means that the chance of observing Zeile's statistic (or one that's more extreme) is higher for the streaky model than for the consistent model. In other words, our streaky model shows a better fit to Zeile's data than does the consistent model.

	Max − Min	Number of long streaks (8 or longer)	Number of runs	Number of 0-hit games	Number of 3+-hit games
Zeile	0.479	2	36	26	7
Streaky 1	0.567	0	37	25	8
Streaky 2	0.373	0	42	41	3
Streaky 3	0.456	1	32	29	8
Streaky 4	0.418	0	33	23	6
Streaky 5	0.394	1	36	24	6
Streaky 6	0.466	0	36	35	6
Streaky 7	0.400	0	40	34	10
Streaky 8	0.339	0	47	19	8
MEAN	0.426	0.25	37.9	28.8	6.9
p-value	0.180	0.18	0.5	0.56	0.43

TABLE 5-8   Statistical Values for Todd Zeile and Eight Simulated Streaky Hitters

	Max − Min	Number of long streaks (8 or longer)	Number of runs	Number of 0-hit games	Number of 3+-hit games
Zeile	0.479	2	36	26	7
Mean Consistent hitters	0.327	0.5	40.4	23.8	5.1
p-value Consistent hitters	0.01	0.06	0.26	0.34	0.23
Mean Streaky hitters	0.426	0.25	37.9	28.8	6.9
p-value Streaky hitters	0.18	0.18	0.5	0.56	0.43

TABLE 5-9   Observed Statistics and *p*-Values of These Statistics for the Consistent and Streaky Hitters

Mr. Consistent or Mr. Streaky?

We have described two probability models for Zeile's hitting data and presented some evidence that the streaky model provides a better description of Zeile's performance. We will now be more specific. Suppose that, before looking at any data, you believe that the Mr. Consistent and the Mr. Streaky models are equally plausible descriptions of Zeile's hitting. After seeing Zeile's data, what do you believe about these two models?

We will illustrate a simple way of computing the following:

Pr(Zeile is a streaky hitter)

To do this, we will use the simulated data of the consistent and streaky hitters and one unusual statistic from Zeile's hitting data. Zeile had seven games in the first half of 1999 where he had three or more hits. Based on Zeile's performance, we consider the two following mutually exclusive events:

"Fewer than seven 3+-hit games"

and

"Seven or more 3+-hit games"

We will compute the probability that Zeile is a streaky hitter based on the second of these: seven or more 3+-hit games.

Suppose that we have 2000 hitters like Todd Zeile, and there are two possible models for his hitting—consistent or streaky. If we think that the chance Zeile is a consistent hitter is the same as the chance that he is a streaky hitter, then

	Less than seven 3+-hit games	Seven or more 3+-hit games	Total
Consistent hitter	•	•	1000
Streaky hitter	•	•	1000
TOTAL	•	•	2000

TABLE 5-10 Table Classifying Hitters by Ability (Consistent or Streaky) and Performance (Fewer than Seven or At Least Seven 3+-Hit Games)

	Less than seven 3+-hit games	Seven or more 3+-hit games	Total
Consistent hitter	770	230	1000
Streaky hitter	570	430	1000
TOTAL	1240	760	2000

TABLE 5-11 Table Classifying Hitters by Ability and Performance with Some Counts Filled In

we'll call 1000 of these hitters consistent and 1000 of them streaky. We put these numbers in Table 5-10.

Earlier, we simulated 1000 hitters from the consistent model and found that 23 percent of them, or 230 hitters, had seven or more 3+-hit games. (That means that $1000 - 230 = 770$ hitters had fewer than seven 3+-hit games.) Similarly, of the 1000 hitters from the streaky model, 43 percent, or 430 hitters, had seven or more 3+-hit games. (So $1000 - 430 = 570$ hitters were in the other category.) We place these values in Table 5-11.

To find the probability that Zeile is a consistent or streaky hitter based on his data, we focus on the column headed "Seven or more 3+-hit games" in the table. There were a total of 760 hitters in our simulation who had a large number of 3+-hit games like Zeile. Of these 760 hitters, 230 (30 percent) were consistent hitters and 430 (70 percent) were streaky hitters. So we can say, on the basis of Zeile's large number of 3+-hits games, the following:

Pr(Zeile is streaky) = .70, Pr(Zeile is consistent) = .30

	Max − Min	Number of long streaks (8 or longer)	Number of runs	Number of 0-hit games	Number of 3+-hit games
Zeile	0.479	2	36	26	7
Pr(streaky)	0.95	0.75	0.66	0.62	0.7
Pr(consistent)	0.05	0.25	0.34	0.38	0.3

TABLE 5-12   Values of Five Interesting Statistics for Todd Zeile and the Probabilities That He Is a Streaky Hitter

The probability (70 percent) that Zeile is streaky is somewhat larger than the initial probability of 50 percent, which means that there is some support for true streakiness in Zeile's data. In a similar fashion, we can use another interesting statistic, such as his large number of "long streaks," to compute the probability that he is streaky based on this statistic and the initial assumption that the models Mr. Streaky and Mr. Consistent are equally likely.

Table 5-12 summarizes these calculations for each one of the five "streaky statistics." With the exception of the MAX − MIN statistic, we see that the probability that Zeile is a streaky hitter (in the way that we have defined it) is in the .6 to .7 range, which is higher than the initial probability of .5. The probability that Zeile is streaky is .95 using the MAX − MIN statistic—this tells us that the big difference between Zeile's best and worst moving average is pretty significant and is more typical of a hitter who is truly streaky.

Team Play

We have spent quite a bit of time analyzing the streaky behavior of a single hitter, and we have found that there is some evidence that Todd Zeile is truly streaky. But are groups of players or teams generally streaky? If we performed the above analysis on all major league players or all teams, would we find that many of them are streaky or possess the hot hand?

To partly answer this question, we will look at the win/loss sequences for all 30 major-league teams in 1998. For each team, we collected the game results for all 162 games (approximately) played that season. Figure 5-11 shows this data for the Anaheim Angels. We see they won their first two games, lost the next three, won the next three, lost the next two, and so on.

Were the Angels streaky in 1998? To begin, when we look at the above sequence we see some interesting patterns. Specifically, we see a large number

Games 1–81:

WWLLLWWWLLWLLWLLWLWWWWWWLWWWWLLLLWLWLWLLWLLWWWWLLLLWLWWWWWWWWWWLWWWWLWWWWLWWWWLWLLWW

Games 82–162:

LLLLLLWLWLLWLLWLLWWLWLWLWLLLLWLLLWWWWWLWLWLWLWWLWWWWWWLLWLWLWLWLWWLWLLLWLLLLLWWLLLWLLW

FIGURE 5-11   Win/loss sequence of the 1998 Anaheim Angels.

WWLLLWWWLLWL LWLLWLWWWWWWLWWWWLLLLWLWLWLLWLLWWWWLLLLWLWWWWWWWWWWLWWWWLWWWWLWWWWLWLLWW

6 wins, 6 losses

FIGURE 5-12   Win/loss sequence with winning fraction for games *1 through 12* displayed.

W WLLLWWWLLWLL WLLWLWWWWWWLWWWWLLLLWLWLWLLWLLWWWWLLLLWLWWWWWWWWWWLWWWWLWWWWLWWWWLWLLWW

wins, 7 losses

FIGURE 5-13   Win/loss sequence with winning fraction for games *2 through 13* displayed.

WW LLLWWWLLWLLW LLWLWWWWWWLWWWWLLLLWLWLWLLWLLWWWWLLLLWLWWWWWWWWWWLWWWWLWWWWLWWWWLWLLWW

wins, 7 losses

FIGURE 5-14   Win/loss sequence with winning fraction for games *3 through 14* displayed.

of wins at the end of the first half of the season (including a winning streak of nine games), and a losing streak of six games at the beginning of the second half. As in the analysis of Zeile's data, one can quantify these clusters of wins and losses by the computation of moving winning fractions. Suppose we use a width of 12 games, which corresponds to about two weeks of games. Then we compute moving winning fractions for all groups of 12 games. To start, we look at the Angel's record in games 1 through 12, which is boxed in Figure 5-12. The Angels won 6 and lost 6 in this period for a winning fraction of .500.

We next look at games 2 through 13. From Figure 5-13, we see the Angels won 5 and lost 7, for a winning fraction of .417.

Next, we look at games 3 through 14;—again we see 5 wins and 7 losses, for a .417 winning fraction.

Suppose that we compute this winning fraction for all groups of 12 games. Figure 5-15 graphs the winning fractions against the mean game number. This graph dramatically shows the periods where the Angels were hot and cold during the season. After an initial lukewarm period, the Angels were hot for a short period (around game 20), and then cold for a period. Then the Angels had an extended hot spell from game 50 to game 80, including a 12-game stretch where they actually had a winning fraction over .90. They followed this long hot stretch with an extended cold spell. They conclude the season with a hot spell and a cold spell.

How can we measure the streakiness that we see in this graph of moving fractions? A simple way is just to compute how far the moving fractions are from the overall season winning percentage. The team went 85–77 in 1998, for a winning fraction of .525, which is the location of the horizontal line in Figure 5-15. We can measure the size of the streakiness by finding the sum:

$$\text{Black} = \text{Sum[distance (moving average} - .525)]}$$

This sum essentially is the size of the black region of the graph, so we will call this statistic "Black." If we see a lot of black in the graph, then the team had a pretty streaky season. For the Angels, we compute the following number:

$$\text{Black} = 21.7$$

How does the streakiness that we see in the Angels' season performance compare to that of other teams? Figures 5-16 and 5-17 display the moving fraction graphs (using a window of 12 games) for all 30 major-league teams in 1998. Looking at these graphs and comparing the sizes of the black areas, we see some teams—such as Anaheim, Baltimore, and Detroit—that appear to have had unusually streaky seasons. Each of these teams has a large chunk of black in its

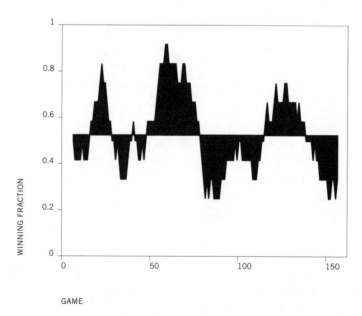

FIGURE 5-15   Moving fraction plot of the winning pattern of the 1998 Anaheim Angels.

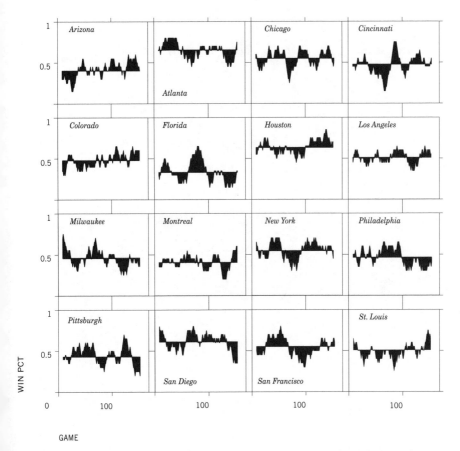

FIGURE 5-16   Moving fraction plots of the winning patterns of National League teams in 1998.

plot, indicating that it had at least one major slump or hot period in its season. Other teams—such as Atlanta, Los Angeles, and Cleveland—appear to have had unusually consistent seasons, since their moving fraction graph stays pretty close to a horizontal line. We can describe the amount of streakiness in each team's graph using the Black statistic. Table 5-13 gives values of Black for all of the teams. The values are consistent with what we saw in Figures 5-16 and 5-17: Anaheim, Baltimore and Detroit are in the 20–27 range, while Atlanta, Los Angeles, and Cleveland are much smaller, in the 11–16 range.

So the 30 major league teams in 1998 appear to vary quite a bit with regard to their consistency across the season, but do these patterns mean anything? We've observed that some teams performed streaky, but that doesn't mean that those

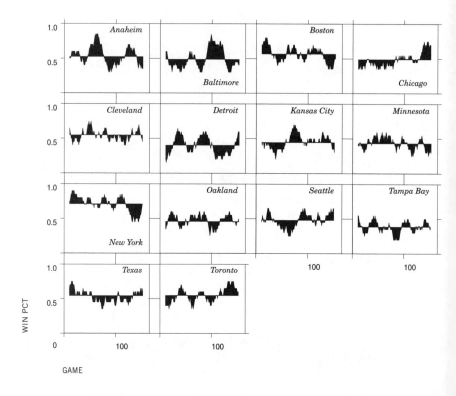

FIGURE 5-17   Moving fraction plots of the winning patterns of American League teams in 1998.

teams actually are streaky. It is possible that they are all consistent teams, but by luck or chance variability, their season performances happened to look streaky.

A Consistent Team

To see if these streaky patterns of team performance mean anything, we propose a few simple models for team abilities and see what type of streaky behavior during a season is predicted based on these models. The simplest model is *The Consistent Team*. This team wins each game during the season with the same probability. It doesn't matter if this team is playing the Yankees or Mets or Devil Rays—the team will always win with the same probability. The winning probabilities of this consistent team are displayed in Figure 5-18 as .525 on game 1, .525 on game 2, .525 on game 3, and so on. Moreover, the results of different games are independent. The chance that our consistent team wins a particular game is unaffected by what happens in previous games. The Consistent Team,

Team	Gray	p-value			
Baltimore	26.6	0.002	Boston	16.3	0.52
Anaheim	21.7	0.04	Chicago White Sox	16.2	0.58
Detroit	21.6	0.05	Montreal	16.1	0.52
Cincinnati	20.7	0.08	Philadelphia	16.1	0.56
Florida	20	0.07	Arizona	15.8	0.56
Pittsburgh	19.9	0.12	Atlanta	15.8	0.52
New York Yankees	18.4	0.09	Oakland	15.6	0.65
San Francisco	18.4	0.25	New York Mets	15.5	0.65
St. Louis	18	0.31	Minnesota	15.4	0.66
Tampa Bay	17	0.38	Colorado	14.9	0.74
Chicago Cubs	16.8	0.48	Kansas City	14.8	0.73
Toronto	16.7	0.47	Texas	14.6	0.77
Milwaukee	16.6	0.5	Cleveland	14.1	0.83
San Diego	16.5	0.47	Houston	13.1	0.88
Seattle	16.4	0.5	Los Angeles	11.9	0.98

TABLE 5-13   Values of the Streaky Statistic Black and the *p*-Value of This Statistic Assuming a Consistent Team

as we have defined it, seems pretty unbelievable–since the chance that a team wins a baseball game clearly depends on a number of different factors—but we'll show that it is a useful model to consider.

Now that we've defined The Consistent Team, we can see how it performs in a 162-game season by using simulation. Let's illustrate how we do this simulation using the Angels as an example. As mentioned earlier, the Angels had a winning fraction of .525 in 1998. Suppose that this team was truly consistent and it won each game with probability .525. We then can play a complete simulated Angels season by using a random spinner 162 times, where the Win region in the spinner is equal to .525. After we simulate the Angels season, we check for streakiness using the moving fraction plot that we've used earlier. We see some black area in the graph, and we measure the size of the streakiness by the statistic Black.

We repeat this simulation for 1000 seasons, and for each we compute the statistic Black. So when we are done, we get 1000 values of the streaky statistic Black. This simulation tells us how much streakiness we will observe in the team performance if the team was truly consistent.

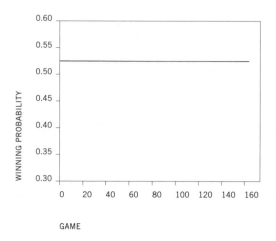

FIGURE 5-18 Graph of the probability of winning across games for The Consistent Team.

In the 1998 Angels season, we observe some streakiness and Black = 21.7. To see if this is unusually large for a consistent team, we compare it to the values of Black in the simulation. We compute a *p*-value which is the chance that the value of Black is at least as large as 21.7, assuming the consistent model. In Table 5-13, we see the following:

$$p\text{-value} = \Pr(\text{Black is at least as large as } 21.7) = .04$$

Since this probability is small, it seems that the 1998 Angels are not behaving like a truly consistent team.

We repeat this simulation for each of the 30 teams. The *p*-values of the observed values of Black are shown in Table 5-13. What is remarkable is that most of the *p*-values are large, and only six of the thirty teams (shown in bold type) have *p*-values under 10 percent. If all 30 teams were truly consistent teams, then one would expect 3 out of the 30 teams to have a *p*-value smaller than 10 percent, and the six observed *p*-values under 10 percent are not much more than what we expect. So we can conclude that the streakiness (black matter) that we observe in the moving fraction plot generally agrees with the observed streakiness of a truly consistent team.

A Streaky Team

We have described what it means for a team to be truly consistent. What does it mean for team to have a streaky nature? We use a notion of streakiness here that is different from what we used for Todd Zeile. We do this to show that there are a number of plausible ways of representing streaky behavior. This model, like the one used earlier, assumes that the winning probability can change across the season. Also, if a team is playing well (or poorly) during a particular game, it is more likely to play well (or poorly) in the next game.

If a team is streaky, we will assume that during the season it can be in one of three possible states, which we will call "hot," "cold," and "lukewarm," or "average." When the team is hot, it wins with a high probability p_H, when it is lukewarm it wins with a smaller probability p_{av}, and when it is cold it wins with the smallest probability p_C. Also we divide the 162-game season into 9 periods of 18 games and assume that during the season the team will be hot for three periods, lukewarm for three periods, and cold for three periods. (Here 18 days is approximately 3 weeks, so we're assuming that a team will remain in the same winning state for about three weeks.) Figure 5-19 shows how the winning probability can change for this type of streaky team. In this graph, the cold, lukewarm, and hot winning probabilities are assumed respectively to be .425, .525, and .625. We see that this particular team is hot for the first 54 games, lukewarm for the next 36, cold for the next 36, lukewarm for the next 18, and cold for the final 18.

Just as before, we use a simulation to see how a streaky team, of the type just

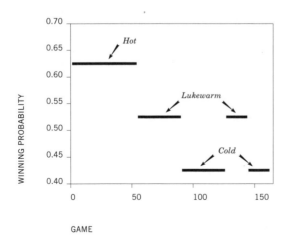

FIGURE 5-19 Graph of the probability of winning across games for a streaky team.

described, will perform during a 162-game season. We illustrate how we do this simulation for Anaheim. In 1998, this team had a winning fraction of .525. We assume that, when Anaheim is hot, it wins with a probability of .625—that is, .1 greater than its season winning fraction. Likewise, it wins with probability .1 lower, or .425, when it is cold. When the team is lukewarm, it wins with probability .525. We divide the season into nine periods, where the team is hot for three periods, cold for three periods, and lukewarm for three periods. (The periods are randomly placed in the season.) After we have decided on the winning probabilities for all games, we simulate game results (wins and losses). Based on a moving fraction plot, we compute the statistic Black.

For each team, we perform this simulation using the streaky team model for 1000 seasons. Each time we do the simulation, we compute the streaky statistic Black. For each of the six teams that appeared streaky, we compute a p-value. This is the probability that the statistic Black is at least as large as the observed value, assuming the streaky model. The results are shown in Table 5-14.

To understand what Table 5-14 is telling us, let's focus on Anaheim. We saw early that Anaheim had a moving fraction plot that looked streaky, and we measure this streakiness by the value Black = 21.7. We saw that this value is a bit unusual for a team that is truly consistent. That is, if a team went through the season always winning with the same probability, then the chance that you would see a Black value this large or larger is only 4 percent. Now, if the team was really streaky (in the way we defined it), we see that the chance of seeing this extreme value of Black is 14 percent, considerably more than 4 percent. So Anaheim's streaky performance is more consistent with a streaky model than a consistent model. Also, if we thought initially that Anaheim was equally likely

Team	Gray	p-value Consistent model	p-value Streaky model	Pr(team is streaky)
Cincinnati	20.7	0.08	0.3	0.79
Florida	20	0.07	0.23	0.77
Anaheim	21.7	0.04	0.14	0.78
Baltimore	26.6	0.002	0.016	0.89
Detroit	21.6	0.05	0.13	0.72
New York Yankees	18.4	0.09	0.37	0.8

TABLE 5-14   Values of the Streaky Statistic Black (p-Values of This Statistic Assume a Consistent Team and a Streaky Team Model, and the Probability the Team Is Streaky)

to be a consistent or streaky team, then Table 5-14 says that the new probability that Anaheim is streaky (given this data) is .78. Looking at the whole table, we see that there is some evidence that each of the teams is truly streaky, and that Baltimore has the strongest evidence of streakiness.

Thinking about Streakiness — Again

What have we learned about streakiness? First, and maybe most important, we understand now (we hope) that there is a difference between *observed streakiness* and *streaky ability*. Every day during the baseball season, we're confronted with interesting observed statistics—say, that Barry Bonds is 0 for 12 in his most recent series of at-bats. Stats like these are just indications that Barry is having a hitting problem. But they are not, after all, much more than statistics in isolation. We're more interested in streaky ability, and what we learn about this ability based on larger collections of data.

In this chapter, we've described several models, such as consistency and streakiness, that tell how an individual's (or team's) hitting (or winning) probability changes over time. For Todd Zeile, we described two models, called Mr. Consistent and Mr. Streaky, and showed that there is some evidence that Zeile was exhibiting streaky behavior. We did a similar analysis for the 1998 teams, and found some evidence that a few of them were truly streaky.

One basic thing we have learned is that it is pretty tough to interpret streakiness data. Even if a hitter like Zeile is really consistent—that is, he gets a hit at each at-bat with the same probability—we can see very interesting patterns. The problem is how to make sense of those patterns. Our old caution—about drawing inferences from small datasets—becomes especially important if you are trying to draw meaningful inferences from the ups and downs of a player or team. One should, at the least, be very cautious in thinking a player is streaky just because an announcer says he was hot last week, and this week he's cold.

There is a related issue that fans should be aware of—namely, *selection bias*. Why did we decide to look at the hitting statistics for Todd Zeile? Well, we had heard through the media that he was a streaky player. In other words, we selected Zeile since his hitting statistics were interesting to look it. Now consider the opposite situation: What if we had heard from a TV broadcaster that Tony Gwynn had gotten 4 hits in his last 12 at-bats? Would we have picked up on this information, and done a statistical study? The answer is a clear No, since this data is not interesting—hitting statistics like these are what we expect from Gwynn, since his lifetime batting average is over .300. It would be interesting to hear, say, that Tony is hitless in his last 20 at-bats, since this data would be far from what we expect. But Tony going 4 for 12? That's old news.

The point we're trying to make is that we look at interesting baseball data and ignore noninteresting data, and that fact alone makes the interesting baseball data appear more significant than it really is. We say that inference from this data is *biased*, or misrepresentative of reality, since the data has been selected exactly because it appears unusual.

To properly decide if hitters or teams are generally streaky, we need to look at a large amount of data that doesn't suffer from selection bias. We did this in the case of the team data—we looked at all teams that played in the 1998 season and didn't select the ones that had interesting win/loss patterns. What we found is that only six of the thirty teams had streaky patterns of wins and losses that did not conform to a consistent (constant-probability) model. Only one team (Baltimore) was unusual with respect to streakiness. Suppose that we look for true streakiness among 30 teams and decide that there is "significance" if the p-value is smaller than 10 percent. Then, even if all of the teams are consistent or non-streaky, we would expect, by chance, that 10 percent of the 30 (or 3 teams) would show "significant" streakiness. So really there is not strong support in this 1998 season for teams displaying streakiness.

There is still an active effort among statisticians to detect streakiness (streaky *ability*) from sports data. Some researchers think that there is little statistical evidence for true streakiness—people just *think* there is streakiness since they are confused by streaky patterns in random sequences. But we'll end this chapter with a note of optimism based on our experience playing sports. Neither of us were good baseball players in Little League, but we both shot a bit of basketball when we were young. On some days, we would feel that we had the right shooting touch and could make any basket we would try. On these days, we believed we had a true hot hand—our shooting stroke had just the right rhythm so that we had a high probability of making a shot. Other days, we would be out of rhythm and have little feeling for the location of the hoop—we would have a much smaller probability of making a shot. So a hot hand refers to a *feeling*—it's an intrinsic characteristic of our shooting ability.

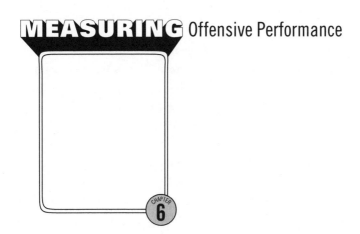

MEASURING Offensive Performance

On April 17, 1960, a trade unique in baseball history was consummated. Cleveland outfielder Rocky Colavito was swapped even up for Detroit outfielder Harvey Kuenn. What made this trade so remarkable was that Colavito was the 1959 American League home-run champion (he hit 42 of them), while Kuenn was the 1959 American League batting champion (with a .353 average). Never before (and never since) have two players been traded for each other just one year after such tremendous personal accomplishments. The trade begged the question, which type of player is more valuable, a great power hitter or a great hitter for average?

The 1959 offensive stats for Colavito and Kuenn are shown in Table 6-1. Colavito had a big edge in home runs, and he received more free passes to first base. Kuenn got more hits overall, and aside from home runs, more extra-base hits. So the situation is complicated: It really is not easy to say that one of these players was better than the other—there are simply too many categories to compare. Nonetheless, the general managers of Cleveland and Detroit must have seen clear-cut (although opposing) advantages, or they would not have made the trade.

Player	AB	H	2B	3B	HR	SH	SF	SB	CS	BB	IBB	HBP
Rocky Colavito	588	151	24	0	42	0	3	3	3	71	8	2
Harvey Kuenn	561	198	42	7	9	3	4	7	2	48	1	1

TABLE 6-1   1959 Offensive Records for Rocky Colavito and Harvey Kuenn

The Great Quest

If there is one great quest in baseball statistics, it is the search for the best formula for evaluating offensive performance. Who is the more valuable player, a Tony Gwynn type of hitter, who has a high batting average but little power? Or a Mike Schmidt type, who displays great power but has a low batting average? And just how valuable is speed in a player? Or the ability to draw a walk? Or . . .

The question of how to quantify offensive performance is a classic statistical problem. Offensive statistics offer a rich but not overwhelming set of dimensions with which to measure players. The question becomes how best to combine these different measurements into a single number that best reflects the offensive value of the player.

We examined the players with the top three career batting averages in each decade. Only players with more than 5000 plate appearances were considered.[1] Each player was assigned to a single decade according to the midpoint of his career. For example, Rogers Hornsby debuted in 1915 and retired after the 1937 season. His career midpoint was 1926, so for our purposes here, he is assigned to the 1920s. (And 1926 was, incidentally, the year Hornsby, as player-manager, led the St. Louis Cardinals to their first World Championship.)

Table 6-2 shows for each decade the total number of players, and the number and percentage of players who had more than 5000 career plate appearances.[2] A total of 5000 plate appearances indicates a substantial major-league career; less than 10 new players per year (on average) achieve this milestone. The low percentage of 1940s players in this category is most likely caused by military service in World War II. The equally low percentage in the 1990s is probably a temporary effect; many players whose mid-career decade will be the 1990s are still active, which means they still have time to reach this 5000 career plate appearance threshold.

---

[1] The sum of at-bats, walks, and hit by pitcher was used here for plate appearances.

[2] The primary data source for this chapter was Sean Lahman's database (now called *The Baseball Archive*), available on the web at www.baseball1.com. The database—actually a set of databases—is of inestimable value to statistical researchers. The two used here are the team database and the player batting database. Several data items (such as Grounded Into Double Play) are unavailable. Others are incomplete: Sacrifice Flies (complete from 1954 on), Caught Stealing (complete from 1951 on), Sacrifice Hits (complete from 1894 on), Hit by Pitcher (complete from 1887 on), and Stolen Bases (complete from 1886 on). Where a data item was unavailable, its value was assumed to be zero. The analyses presented here were performed initially with Version 2.2 of the database, which covered all seasons through 1998. The analyses were extended to include the 1999 season when Version 3.0 of the database was made available.

Decade	Total	>5000	%>5000
1880s	899	29	3.2%
1890s	789	47	6.0%
1900s	1013	39	3.8%
1910s	1558	58	3.7%
1920s	1208	54	4.5%
1930s	960	58	6.0%
1940s	1222	46	3.8%
1950s	1046	54	5.2%
1960s	1144	71	6.2%
1970s	1306	100	7.7%
1980s	1399	96	6.9%
1990s	2248	87	3.9%

TABLE 6-2   Total Players from Each Decade and Number and % with More Than 5000 Plate Appearances

Table 6-3 lists the leading players in each decade according to batting average. Scanning this list, we see only the names of very good hitters. All players prior to 1980 are in the Hall of Fame, with the following exceptions: Browning, Jackson, Alou, Garr, and Oliver. Puckett,[3] Molitor, Gwynn, and Boggs are destined for the Hall, and Frank Thomas is likely to make it as well. But were they truly the best of their period in producing runs? For example, Richie Ashburn is a personal favorite of ours, but even we doubt that he was a greater hitter than his contemporary Mickey Mantle. And was Paul Waner a better hitter than Jimmie Foxx? And say Hey! Where's Willie Mays?

Over the years, many systems have been offered to ascertain the offensive value of players. And each year, more are added to the list by sportswriters and fans. *The 1999 Big Bad Baseball Annual* alone listed over 20 systems for evaluating offensive performance. What's going on here? Isn't the tried and true Batting Average enough?

As it turns out, the answer is No. Batting average actually has a relatively poor correlation with runs scored. Interestingly, the best way to gauge the value of systems for rating *individual* offensive players is to analyze *team* data. In

[3] Kirby Puckett was enshrined in 2001.

Decade	Player	AVG					
1880s	Pete Browning	.341	Cap Anson	.333	Roger Connor	.317
1890s	Ed Delahanty	.346	Billy Hamilton	.344	Dan Brouthers	.342
1900s	Willie Keeler	.341	Nap Lajoie	.338	Honus Wagner	.327
1910s	Ty Cobb	.366	Joe Jackson	.356	Tris Speaker	.345
1920s	Rogers Hornsby	.358	Babe Ruth	.342	Harry Heilmann	.342
1930s	Lou Gehrig	.340	Al Simmons	.334	Paul Waner	.333
1940s	Ted Williams	.344	Joe Dimaggio	.325	Joe Medwick	.324
1950s	Stan Musial	.331	Jackie Robinson	.311	Richie Ashburn	.308
1960s	Roberto Clemente	.317	Matty Alou	.307	Hank Aaron	.305
1970s	Rod Carew	.328	Ralph Garr	.306	Al Oliver	.303
1980s	Kirby Puckett	.318	Don Mattingly	.307	Paul Molitor	.306
1990s	Tony Gwynn	.339	Wade Boggs	.328	Frank Thomas	.320

TABLE 6-3   Players with Highest Career Batting Average from Each Decade

baseball, a single individual is rarely responsible for production of a run. Batters get on base, then other batters advance them. The "offensive credit," as it were, is shared—by the runner who got on base and scored, the players who advanced the runner, and the player who batted him in for the RBI. The fact that scoring is a series of events involving more than a single player is one reason why the standard counts for runs scored and runs batted in are not satisfactory evaluators of individuals. Even the sum of total runs scored plus RBIs minus HRs has not achieved widespread use.

Over the next few pages, we'll take a look at some of the most widely used stats for offensive performance, analyze how closely they align with runs produced, examine what they tell us, and what they don't tell us, about the run-producing value of noted players.

Runs Scored per Game

Let's start our investigation by looking at the runs scored per game by teams in 1998 as presented in Table 6-4. Runs per game (R/G) ranged from the low of 3.827 by the Tampa Bay Devil Rays to the high of 5.957 by the World Champion New York Yankees. (Coincidentally, both teams were in the Eastern Division of the American League.) This represents quite a spread. Over the course of the 1998 season, the Yankees scored more than 50 percent more runs per game than the Devil Rays.

Team	Runs per Game		
New York Yankees	5.957	Anaheim Angels	4.858
Texas Rangers	5.802	Cincinnati Reds	4.630
Boston Red Sox	5.407	San Diego Padres	4.623
Houston Astros	5.395	Minnesota Twins	4.531
Seattle Mariners	5.335	Detroit Tigers	4.457
Chicago White Sox	5.315	Kansas City Royals	4.435
Cleveland Indians	5.247	Philadelphia Phillies	4.401
San Francisco Giants	5.184	Milwaukee Brewers	4.364
Atlanta Braves	5.099	New York Mets	4.358
Colorado Rockies	5.099	Los Angeles Dodgers	4.130
Chicago Cubs	5.098	Florida Marlins	4.117
Baltimore Orioles	5.043	Arizona Diamondbacks	4.105
Toronto Blue Jays	5.037	Pittsburgh Pirates	4.012
St. Louis Cardinals	5.000	Montreal Expos	3.975
Oakland A's	4.963	Tampa Bay Devil Rays	3.827

TABLE 6-4   Runs per Game for Major League Teams in 1998

Now let's suppose you had absolutely no knowledge of baseball except for the information in Table 6-4 (minus the team names), and that you are asked to guess how many runs per game a certain team scored in 1998. What would be your best guess? You wouldn't pick the highest or lowest value, since this would make your possible error very large. Most likely you would pick some value in the middle of the distribution. Perhaps you would calculate the average (or mean) of all the values (4.794 runs per game) and use it as your best guess.

In fact, the average *is* the best guess (or estimate) you can make . . . without any further information, that is. Table 6-5 shows the runs per game values, the guess based on the average, and the difference between this guess and the actual value. We will refer to this difference as the *Error* in the estimate. At the bottom of the Error column is the average value of the errors, 0. Basically, what this means is that if you use the average value as your guess, you will overestimate as much as you underestimate in repeated guesses.

Another column presents the square of the error; that is, *Error* × *Error*. Doing this has a great advantage. Now each error, whether positive or negative, has been converted to a positive value. So, an error of –1 (overestimating by 1 run

Team	Runs per Game	Estimate	Error	Error × Error
New York Yankees	5.957	4.794	1.163	1.352
Texas Rangers	5.802	4.794	1.008	1.017
Boston Red Sox	5.407	4.794	.613	.376
Houston Astros	5.395	4.794	.601	.361
Seattle Mariners	5.335	4.794	.541	.293
Chicago White Sox	5.315	4.794	.521	.271
Cleveland Indians	5.247	4.794	.453	.205
San Francisco Giants	5.184	4.794	.390	.152
Atlanta Braves	5.099	4.794	.305	.093
Colorado Rockies	5.099	4.794	.305	.093
Chicago Cubs	5.098	4.794	.304	.093
Baltimore Orioles	5.043	4.794	.249	.062
Toronto Blue Jays	5.037	4.794	.243	.059
St. Louis Cardinals	5.000	4.794	.206	.042
Oakland A's	4.963	4.794	.169	.029
Anaheim Angels	4.858	4.794	.064	.004
Cincinnati Reds	4.630	4.794	−.164	.027
San Diego Padres	4.623	4.794	−.171	.029
Minnesota Twins	4.531	4.794	−.263	.069
Detroit Tigers	4.457	4.794	−.337	.114
Kansas City Royals	4.435	4.794	−.359	.129
Philadelphia Phillies	4.401	4.794	−.393	.154
Milwaukee Brewers	4.364	4.794	−.430	.185
New York Mets	4.358	4.794	−.436	.190
Los Angeles Dodgers	4.130	4.794	−.664	.441
Florida Marlins	4.117	4.794	−.677	.458
Arizona Diamondbacks	4.105	4.794	−.689	.475
Pittsburgh Pirates	4.012	4.794	−.782	.611
Montreal Expos	3.975	4.794	−.819	.670
Tampa Bay Devil Rays	3.827	4.794	−.967	.935
Average	4.794	4.794	.000	.300

TABLE 6-5   Predicting Runs per Game for MLB Teams in 1998 (Prediction = Average)

per game) is treated the same as an error of 1 (underestimating by 1 run per game). The average of the squared errors, or *Mean Squared Error (MSE),* is presented at the bottom of the column. The square root of MSE, or *Root Mean Squared Error (RMSE)*, an estimate of the standard deviation of the error distribution, provides a measure of how much you may expect to overestimate or underestimate in your guesses. Approximately two-thirds of all errors are between –RMSE and +RMSE runs per game. Here the RMSE is the square root of .300, which equals .548 runs per game. From the Error column in Table 6-5, we see that 20 teams (exactly two-thirds of 30 teams) have estimates with errors between –.548 and .548.

One can show mathematically that the average guess results in the lowest possible MSE among all other possible guesses. Let's demonstrate this point with an example. Suppose you had guessed 5 instead of 4.794. How would the MSE have changed? Table 6-6 is exactly the same as Table 6-5 except that 5 has been substituted for 4.794 as the estimate. The Error column has been calculated the same way, by subtracting the estimate from actual runs per game. The first change we notice is that the average error is not 0 anymore. It's –.206, reflecting the tendency of 5 to overestimate the runs per game for a team. The MSE value (the average of Error × Error) is .342, which is greater than the MSE value .300 in Table 6-5, when the average 4.794 was used as the estimate. Since RMSE is just the square root of MSE, the RMSE value is greater as well (.585 versus .548 runs per game).

Figure 6-1 shows the RMSE value for every reasonable guess of runs scored per game. For each guess (shown on the *x*-axis), we followed the same procedure as in Table 6-6 (where 5 was the guess). That is, using the guess as the estimate, we subtracted it from each data value to obtain an error; the errors were squared and the average value of their squares calculated. The square root of the average gave us the RMSE for that guess. The plot shows the RMSE for each guess. The RMSE values for our guesses of 4.794 and 5 runs per game are included in the line. For example, our original guess was 4.794 runs per game based on the runs scored per game averaged over all teams. The calculation in Table 6-5 gave a RMSE of .548 runs per game if this value were used as an estimate. Figure 6-1 displays a dot on the curve where the RMSE result of this guess is plotted. Another dot shows the RMSE value when 5 runs per game is the guessed estimate. Note that this dot is higher than that for the 4.794 guess because the RMSE for a guess of 5 runs per game is higher. In fact, the RMSE reaches its lowest level with the 4.794 runs per game guess. Clearly, the RMSE is smallest when the guess is based on the average runs scored per game.

Team	Runs per Game	Estimate	Error	Error × Error
New York Yankees	5.957	5	.957	.915
Texas Rangers	5.802	5	.802	.644
Boston Red Sox	5.407	5	.407	.166
Houston Astros	5.395	5	.395	.156
Seattle Mariners	5.335	5	.335	.112
Chicago White Sox	5.315	5	.315	.099
Cleveland Indians	5.247	5	.247	.061
San Francisco Giants	5.184	5	.184	.034
Atlanta Braves	5.099	5	.099	.010
Colorado Rockies	5.099	5	.099	.010
Chicago Cubs	5.098	5	.098	.010
Baltimore Orioles	5.043	5	.043	.002
Toronto Blue Jays	5.037	5	.037	.001
St. Louis Cardinals	5.000	5	.000	.000
Oakland A's	4.963	5	−.037	.001
Anaheim Angels	4.858	5	−.142	.020
Cincinnati Reds	4.630	5	−.370	.137
San Diego Padres	4.623	5	−.377	.142
Minnesota Twins	4.531	5	−.469	.220
Detroit Tigers	4.457	5	−.543	.295
Kansas City Royals	4.435	5	−.565	.319
Philadelphia Phillies	4.401	5	−.599	.359
Milwaukee Brewers	4.364	5	−.636	.404
New York Mets	4.358	5	−.642	.412
Los Angeles Dodgers	4.130	5	−.870	.758
Florida Marlins	4.117	5	−.883	.779
Arizona Diamondbacks	4.105	5	−.895	.801
Pittsburgh Pirates	4.012	5	−.988	.975
Montreal Expos	3.975	5	−1.025	1.050
Tampa Bay Devil Rays	3.827	5	−1.173	1.376
Average	4.794	5	−.206	.342

TABLE 6-6   Predicting Runs per Game for MLB Teams in 1998 (Prediction = 5 Runs/Game)

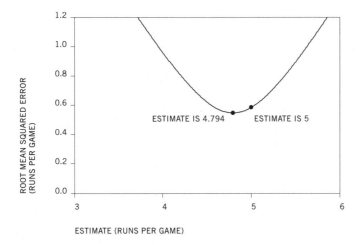

FIGURE 6-1   Root Mean Squared Error (RMSE) values for different guesses of team runs scored per game in 1998.

Batting Average and Runs Scored per Game

We have established that, without any knowledge about the teams, the best prediction of run production is a blind guess based on the average runs scored per game. But what if more information is available? How much can we improve on our guess?

We will examine this with respect to the most popular measure of offensive production, the *Batting Average (AVG)*. Batting Average is the most quoted of all baseball statistics in the print and broadcast media; it is simply the number of hits divided by the number of at-bats. But is AVG worthy of its standing as the number-one baseball stat for individual offensive performance? As the ratio of two easily obtained quantities, it has the strength of simplicity. AVG also has intuitive appeal. It seems reasonable that greater production of hits would lead to greater production of runs. But how strong is this relationship?

Figure 6-2 plots runs scored per game versus AVG for all teams in the 1998 season. As expected, the plot shows a strong correlation between the two measures. Teams with high AVGs tend to have high run production, and teams with low AVGs tend to have low run production. Still, note that the New York Yankees had the highest run productivity without having the highest AVG. On the other hand, the Tampa Bay Devil Rays had the lowest run productivity and yet were far from being the worst team in AVG.

We have drawn a line through the dot cloud. This line is special in the sense that it has the *lowest* RMSE of any other possible line through the plot. The

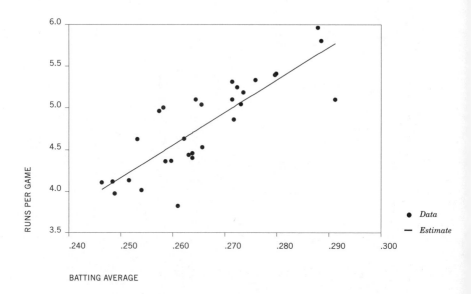

FIGURE 6-2   Runs per Game versus Batting Average for 1998 MLB Teams. (Dots represent actual data, line represents AVG Line.)

RMSE for this line is calculated using the same technique used in Tables 6-5 and 6-6. Here the estimate based on a single average value is replaced by the value of the line for the appropriate AVG:

Estimated Team Runs per Game $= -5.592 + (39.03 \times AVG)$

Let's use this *AVG Line* to predict the run production for the New York Yankees. The Yankees' 1998 AVG was .288, and so we predict the Runs per Game to be as follows:

Estimated Team Runs per Game $= -5.592 + (39.03 \times .288) = 5.646$

This prediction is less than the actual Yankee rate of 5.957 Runs per Game. The Yankees scored more frequently than we might expect based on the Team Batting Average.

Table 6-7 presents the predictions for all 1998 teams based on the Team Batting Average as well as the calculations of the MSE for these predictions. It has several similarities with Table 6-5. In both, the average of the estimates equals 4.794, which is the average runs scored per game over all teams. Also, the average of the errors is 0 in both tables. However, the big difference is in the

Team	Runs per Game	Estimate	Error	Error × Error
New York Yankees	5.957	5.646	.310	.096
Texas Rangers	5.802	5.672	.131	.017
Boston Red Sox	5.407	5.334	.074	.005
Houston Astros	5.395	5.325	.070	.005
Seattle Mariners	5.335	5.177	.158	.025
Chicago White Sox	5.315	5.001	.313	.098
Cleveland Indians	5.247	5.040	.207	.043
San Francisco Giants	5.184	5.087	.097	.009
Atlanta Braves	5.099	5.004	.094	.009
Colorado Rockies	5.099	5.772	−.674	.454
Chicago Cubs	5.098	4.729	.369	.136
Baltimore Orioles	5.043	5.068	−.024	.001
Toronto Blue Jays	5.037	4.773	.264	.070
St. Louis Cardinals	5.000	4.484	.516	.266
Oakland A's	4.963	4.453	.510	.260
Anaheim Angels	4.858	5.014	−.156	.024
Cincinnati Reds	4.630	4.640	−.011	.000
San Diego Padres	4.623	4.289	.334	.112
Minnesota Twins	4.531	4.779	−.248	.061
Detroit Tigers	4.457	4.702	−.245	.060
Kansas City Royals	4.435	4.675	−.240	.058
Philadelphia Phillies	4.401	4.705	−.304	.092
Milwaukee Brewers	4.364	4.543	−.179	.032
New York Mets	4.358	4.501	−.143	.020
Los Angeles Dodgers	4.130	4.227	−.098	.010
Florida Marlins	4.117	4.105	.012	.000
Arizona Diamondbacks	4.105	4.024	.081	.006
Pittsburgh Pirates	4.012	4.319	−.307	.094
Montreal Expos	3.975	4.120	−.144	.021
Tampa Bay Devil Rays	3.827	4.595	−.768	.590
Average	4.794	4.794	.000	.089

TABLE 6-7 Predicting Runs per Game for MLB Teams in 1998 (Estimate = AVG Line)

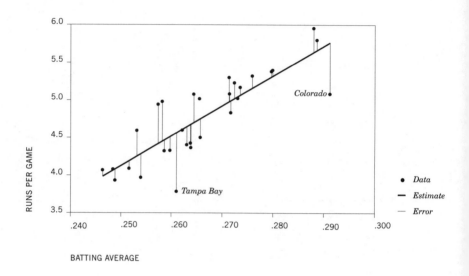

FIGURE 6-3   Runs per Game versus Batting Average for 1998 MLB teams. (Dots represent data, line represents AVG Line; vertical lines represent Error.)

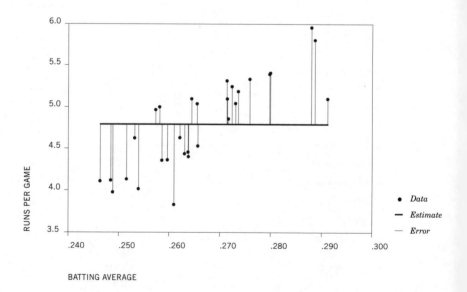

FIGURE 6-4   Runs per Game versus Batting Average for 1998 MLB teams. (Dots represent data, line represents overall average runs per game, vertical lines represent Error.)

MSE and RMSE values. The AVG Line produces a MSE value of .089 in Table 6-7. After taking the square root, this translates to a RMSE value of .299 Runs per Game, almost half the RMSE value for the blind guess estimate in Table 6-5.

This should not come as a surprise. The AVG Line is the line with the best fit to the data. The errors in Table 6-5 were based on an estimate which ignored any team information and estimated a single run production value of 4.794 Runs per Game for all teams. Using the information derived from the Batting Average, we can now estimate team run production correctly within .3 Runs per Game for two-thirds of the teams; before, our estimates were correct within .6 Runs per Game for two-thirds of the teams. Batting Average has thus allowed us to shrink considerably the error in our estimates.

Figure 6-3 is the same as Figure 6-2, but with the addition of a vertical line showing the error of using the AVG Line to predict run production for each team. The figure gives a visual demonstration of the effectiveness of the AVG Line in predicting team run production. The two greatest errors are for the Tampa Bay Devil Rays and the Colorado Rockies. Both teams were expected to score more runs on the basis of Team Batting Average.

Figure 6-4, which is similar to Figure 6-3, shows the effectiveness of prediction on the basis of the overall average Runs per Game. A guess based on this average makes no use of any additional information about the teams, so the prediction is a flat line at the average value. The lengths of the error bars in Figure 6-3 are generally shorter than those in Figure 6-4, which shows why the AVG Line is a better fit to the data and has a lower RMSE.

Up to this point, the Batting Average looks pretty good: it appears to be correlated with team run production. When used in the AVG Line formula, it improves estimation of run production over blindly guessing at the runs scored with the average. But while this is a good start, AVG really hasn't been tested. What if we looked at some other batting measures? Would they do any better?

Slugging Percentage and On-Base Percentage

The *Slugging Percentage (SLG)* is another standard measure of individual baseball hitting performance. Batting Average counts each hit equally, whereas SLG weights each hit according to the number of bases attained:

$$SLG = \frac{1B + (2 \times 2B) + (3 \times 3B) + (4 \times HR)}{AB} = \frac{H + 2B + (2 \times 3B) + (3 \times HR)}{AB}$$

To put it another way, SLG is a measure of total bases achieved divided by at-bats.

The use of the term "percentage" in Slugging Percentage is a misnomer. While SLG is typically less than 1, we see that it is possible for it to be greater than 1, and possibly as high as 4 if every at bat produces a home run. SLG is best understood as either of the following:

- a Rate—the rate at which bases are produced per at-bat

- an Expectation—the expected or average number of bases produced per at-bat

With its emphasis on extra-base hits, SLG improves the rankings of power hitters over high-average "banjo" (non-power) hitters. At least, the "slugging" part of the name is very apt. Table 6-8 presents the players with the highest Slugging Percentage in each decade. The table has much in common with Table 6-3, which simply lists those with the highest Batting Average. But the ordering has changed, and many new players have appeared.

- Harry Stovey replaces Cap Anson in the 1880s.

- Sam Thompson replaces Billy Hamilton in the 1890s.

- Sam Crawford replaces Wee Willie Keeler in the 1900s.

- Hack Wilson replaces Harry Heilmann in the 1920s.

- Jimmie Foxx and Hank Greenberg replace Al Simmons and Paul Waner in the 1930s.

- Johnny Mize replaces Joe Medwick in the 1940s.

- Mickey Mantle and Ralph Kiner replace Jackie Robinson and Richie Ashburn in the 1950s.

- Willie Mays and Frank Robinson replace Roberto Clemente and Matty Alou in the 1960s.

- The 1970s and 1980s saw a complete overhaul; Carew, Garr, and Oliver were swept away by Allen, Stargell, and Jackson, while Schmidt, Rice, and Brett replaced Puckett, Mattingly, and Molitor.

- The 1990s had only Frank Thomas as a repeat; Belle and McGwire replaced Gwynn and Boggs.

Table 6-8, when compared with Table 6-3, also has a large number of present and future Hall of Famers; only Browning, Joe Jackson, Stovey, and Allen are not in the Hall of Fame among the pre-1980s players.

Two principles appear from the comparison of Tables 6-8 and 6-3:

Decade	Player	SLG					
1880s	Roger Connor	.486	Pete Browning	.467	Harry Stovey	.461
1890s	Dan Brouthers	.519	Ed Delahanty	.505	Sam Thompson	.505
1900s	Nap Lajoie	.467	Honus Wagner	.466	Sam Crawford	.452
1910s	Joe Jackson	.517	Ty Cobb	.512	Tris Speaker	.500
1920s	Babe Ruth	.690	Rogers Hornsby	.577	Hack Wilson	.545
1930s	Lou Gehrig	.632	Jimmie Foxx	.609	Hank Greenberg	.605
1940s	Ted Williams	.634	Joe Dimaggio	.579	Johnny Mize	.562
1950s	Stan Musial	.559	Mickey Mantle	.557	Ralph Kiner	.548
1960s	Willie Mays	.557	Hank Aaron	.555	Frank Robinson	.537
1970s	Dick Allen	.534	Willie Stargell	.529	Reggie Jackson	.490
1980s	Mike Schmidt	.527	Jim Rice	.502	George Brett	.487
1990s	Mark McGwire	.587	Albert Belle	.573	Frank Thomas	.573

TABLE 6-8   Batters With Highest Career Slugging Percentage from Each Decade

- *Power hitters replace singles hitters in the SLG ratings.* The players who are in both tables were power hitters who also hit for average.

- *The differences between the two tables appear to be greater in recent years.* This suggests that in the past, power hitters were also great hitters for average, while recent hitters are more apt to be good power hitters *or* good hitters for average, not both.

Figure 6-5 plots Team Runs per Game versus Slugging Percentage. As in Figures 6-3 and 6-4, we have also plotted the SLG Line and the errors for each data point. The formula for the SLG Line is as follows:

Estimated Team Runs per Game $= -2.135 + (16.50 \times SLG)$

Another popular measure for hitting performance is the *On-Base Percentage (OBP)* which we examined extensively in Chapter 2. Recall that OBP is defined as follows:

$$OBP = \frac{H + BB + HBP}{AB + BB + HBP + SF}$$

OBP is used as an estimate of the probability of getting on base in a plate appearance. Table 6-9 lists, by decade, the batters with the highest career OBPs. Again, aside from re-ordering players who appeared in the AVG and SLG lists, a

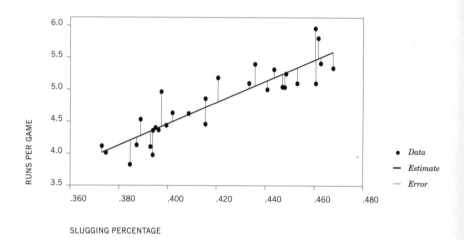

FIGURE 6-5   Runs per Game versus Slugging Percentage for 1998 MLB teams. (Dots represent data, line represents SLG Line, vertical lines represent Error.)

Decade	Player	OBP					
1880s	Pete Browning	.403	Roger Connor	.397	Cap Anson	.393
1890s	Billy Hamilton	.455	Dan Brouthers	.423	Cupid Childs	.416
1900s	Roy Thomas	.413	Honus Wagner	.391	Hughie Jennings	.390
1910s	Ty Cobb	.433	Tris Speaker	.428	Eddie Collins	.424
1920s	Babe Ruth	.474	Rogers Hornsby	.434	Max Bishop	.423
1930s	Lou Gehrig	.447	Jimmie Foxx	.428	Mickey Cochrane	.419
1940s	Ted Williams	.482	Eddie Stanky	.410	Arky Vaughan	.406
1950s	Mickey Mantle	.421	Stan Musial	.417	Jackie Robinson	.409
1960s	Frank Robinson	.389	Willie Mays	.384	Eddie Mathews	.376
1970s	Mike Hargrove	.396	Rod Carew	.393	Joe Morgan	.392
1980s	Rickey Henderson	.405	Tim Raines	.385	Keith Hernandez	.384
1990s	Frank Thomas	.440	Edgar Martinez	.426	Jeff Bagwell	.416

TABLE 6-9 Batters with Highest Career On-Base Percentage from Each Decade

number of new players have entered the picture. Cupid Childs in the 1890s, Roy Thomas and Hughie Jennings in the 1900s, Eddie Collins in the 1910s, Max Bishop in the 1920s, Mickey Cochrane in the 1930s, Eddie Stanky and Arky Vaughan in the 1940s, and Eddie Mathews in the 1960s. Once again we see entire new triplets of players for the 1970s and 1980s, while only Frank Thomas remains in the 1990s.

Some of these new players are in the Hall of Fame (Jennings, Collins, Cochrane, Vaughan, Mathews, and Morgan), some will probably make it there (Henderson), and some may be worthy of consideration (Bishop, Hernandez, and Raines). But do we really think that Mike Hargrove was the best offensive player of the 1970s? Or that Eddie Stanky was the second best player of the 1940s? OBP seems to recognize some excellent players overlooked by AVG and SLG, but it also produces some strange rankings.

Figure 6-6 plots Team Runs per Game versus On-Base Percentage. We have also plotted the OBP Line and the errors for each data point. The formula for the OBP Line is:

$$\textit{Estimated Team Runs per Game} = -7.273 + (36.03 \times OBP)$$

In order to compare how well AVG, SLG, and OBP predict 1998 Team Runs per Game, we took the errors in their predictions from Figures 6-3, 6-5, and 6-6 and then sorted them. Table 6-10 presents these results. We see that AVG had

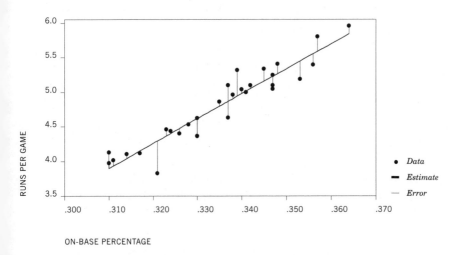

FIGURE 6-6   Runs per Game versus On-Base Percentage for 1998 MLB teams. (Dots represent data, line represents OBP Line, vertical lines represent Error.)

the largest single error (overestimating Tampa Bay's Runs per Game by .768). In fact, AVG had four estimates which were off by more than any estimate based on OBP, which was never off by more than .466 Runs per Game in its predictions. SLG also had two estimates worse than any estimates from OBP. *So* . . . Batting Average is not looking as good as it initially did, and both SLG and OBP seem to do a better job. But we cannot judge by several extremes. We have to examine the entire distribution of errors for AVG, SLG, and OBP.

One way to get a visual perspective on the spread of errors for the different models is to graph side-by-side boxplots of the distributions. We do this in Figure 6-7. If we examine the SLG boxplot, we see the entire extent of the distribution of SLG errors in predicting 1998 Team Runs per Game. The distribution ranges from underestimating one team's run production by .544 Runs per Game up to overestimating another team's run production by .385 Runs per Game. You may recall from Chapter 2 that the box in the center of the plot represents the central half of the distribution; 50 percent of the SLG errors fall within this box, 25 percent above the box, and 25 percent below it. The line in the middle of the box represents the central point (or median) of the entire distribution, with 50 percent above the line and 50 percent below it.

A good predictor of Runs per Game results in a tight distribution of errors, one which limits the size of the error in either direction. Viewing the boxplots in Figure 6-7, the better predictor is one with a more limited range (the difference between the highest and lowest values) and a narrower box. Visual inspection of Figure 6-7 indicates that SLG appears to have less spread in its errors than AVG, and OBP has less spread in its errors than SLG.

The results are best summarized by looking at the Root Mean Squared Error (RMSE) for each predictor.

AVG	SLG	OBP
.516	.544	.374
.510	.497	.233
.369	.381	.229
.334	.342	.213
.313	.323	.178
.310	.253	.142
.264	.139	.115
.207	.134	.092
.158	.133	.080
.131	.102	.079
.097	.084	.064
.094	.022	.061
.081	.019	.060
.074	-.004	.058
.070	-.015	.049
.012	-.019	.034
-.011	-.030	.017
-.024	-.036	.006
-.098	-.089	-.013
-.143	-.121	-.014
-.144	-.136	-.031
-.156	-.194	-.072
-.179	-.216	-.131
-.240	-.238	-.159
-.245	-.240	-.186
-.248	-.244	-.240
-.304	-.261	-.253
-.307	-.364	-.259
-.674	-.381	-.262
-.768	-.385	-.466

TABLE 6-10   Errors in 1998 Team Runs per Game for AVG, SLG, and OBP Lines

The RMSE values in Table 6-11 as well as their visual counterparts, the boxplots in Figure 6-7, all indicate that OBP was the superior measure of batting ability leading to runs in 1998. This is a somewhat surprising result, given the explosion of home runs in that epochal year.

What about other years? Could this have been true only in 1998, or does it hold in other years as well? We performed the same analysis for each season since 1876, the inaugural year of the National League. In each year, we found the AVG Line, SLG Line, and OBP Line with respect to predicting Team Runs per Game in that season and calculated the RMSE for each of the three lines. We then found which Line (AVG, SLG, or OBP) had the lowest RMSE (that is, the best fit to Team Runs per Game). Figure 6-8 plots this minimal RMSE value and indicates which measure (AVG, SLG, or OBP) generated it.

We see that the RMSE of the best fit among the three models shows great variability from season to season. In 1948, Slugging Percentage provided the best fit (RMSE = .32 Runs per Game); thus, AVG and OBP both had RMSEs greater than .32. SLG also provided the best fit in 1968, where RMSE is .12 Runs per Game. In that year, scoring was very low; the reduced variability in runs scored may have contributed to the improved capability of SLG to predict run production.

Batting Average	Slugging Percentage	On-Base Percentage
.299	.248	.178

TABLE 6-11   Root Mean Squared Error for 1998 Team Runs per Game Estimates

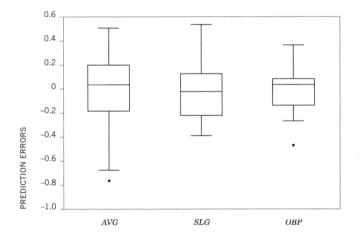

FIGURE 6-7   Boxplots of 1998 Team Runs per Game Errors for AVG, SLG, and OBP.

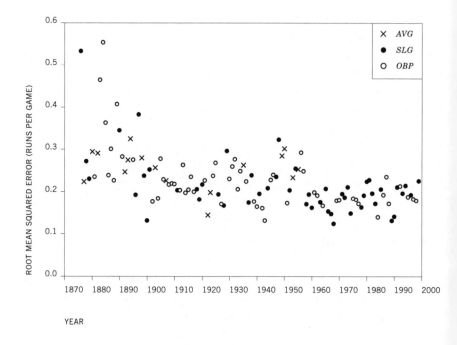

FIGURE 6-8   RMSE of fit to Team Runs per Game since 1876. (Minimum RMSE from AVG, SLG, and OBP.)

Another interesting feature of Figure 6-8 is the improved consistency in fit over time, especially in post-1960 seasons compared to pre-1960 seasons. From 1958 on, every season had a measure with an RMSE value less than .24 Runs per Game. On the other hand, in 35 of the first 82 seasons, the best fit had RMSEs greater than .24. Contributing to this effect may have been the gradual expansion of Major League Baseball from 16 teams in 1960 up to the current 30 teams. Also contributing to this effect are the short seasons (fewer than 100 games) in the early years of professional baseball, which allowed greater variability in run production from the basic hitting events.

For our purposes, the most interesting feature is the relatively few seasons in which Batting Average had the best fit to team run production. AVG has not provided the best fit to run production since 1955. If all three measures (AVG, SLG, and OBP) were equally capable of predicting Team Runs per Game, we would expect AVG to have the best fit (lowest RMSE) in about one-third of the seasons.

In fact, AVG was best in only 16 of 124 seasons, while SLG and OBP split the remaining seasons almost equally (50 and 58 seasons, respectively). The probability of this happening if AVG, SLG, or OBP were equally effective measures is less than 1 in 5 *million*![4]

It is not clear at this point whether Slugging Percentage or On-Base Percentage is the best predictor of run production, but it is abundantly clear that Batting Average is the worst of the three. So the king is dead! Long live the king! But which measure is the new king? SLG and OBP are viable candidates, but why should we restrict ourselves to these choices alone? Can other models provide us with even greater improvements in predicting run production? And if so, how can they be applied to our ultimate goal, evaluating individual players?

Intuitive Techniques

It was not long after the end of World War II that dissatisfaction with the basic crop of baseball offensive performance measures (AVG, OBP, and SLG) initiated research into alternatives. Activity started slowly in the 1950s and really accelerated in the 1960s. Baseball fans must have been thinking, "Hey, we've split the atom! We're sending men to the moon! There's got to be a better way of measuring ballplayers!" Whether this internalized outcry was real or not, activity peaked in the early 1980s, not coincidentally with the creation of the Society of American Baseball Research (SABR). Since then, development has stabilized around several well-established measures.

By far the most popular group of techniques falls into a category we call *intuitive*. With intuitive techniques, no rigorous statistical model is used. (We'll get to those in Chapters 7 and 8.) Instead, the intuitive researcher relies on a vision or paradigm for the workings of baseball, and, inspired by the standard MLB statistics we've talked about earlier in this chapter, "mixes and matches" them to more accurately reflect his or her sense of the game.

The three measures recognized by Major League Baseball discussed earlier all had their origins as intuitive techniques. To develop the Batting Average, it was not necessary to perform analysis of reams of data or develop probabilistic models simulating baseball games. It arose out of a common-sense understanding of baseball. To put it plainly, it makes sense. The other official MLB offensive statistics, OBP and SLG, were developed from a similar intuitive sensibility.

---

[4] This value was calculated as the probability of 16 successes or fewer in 124 trials for a binomial distribution with a probability of success in each trial equal to 1/3.

Fans of the game used these official measures as a starting point. All three had a role to play in the intuitive class of new statistical techniques. These new developments run the gamut from simple tweaks of the standard existing measures to major recombinations of the standard batting data. What we'll see in the balance of this chapter is how AVG, SLG, and OBP can be combined in ways that create paradigms of the game that are closer to what actually happens on the field.

On-Base Plus Slugging (OPS)

Given the relative parity between On-Base Percentage and Slugging Percentage as estimators of team run production, perhaps combining the two would prove to be a useful predictor. This was not the genesis of a model called *On-Base Plus Slugging (OPS)*, but perhaps it provides a reasonably simple explanation for its effectiveness:

$$OPS = OBP + SLG$$

Actually, OPS was developed by Pete Palmer as an easily calculated approximation to his more detailed Linear Weights model (to be covered in Chapter 7). Table 6-12 presents the leading OPS batters in each decade. Most of these players appeared on the SLG list or the OBP list. However, there are some new players, like Jack Clark and Reggie Smith, who struck a good balance of power and getting on-base, but were not leaders in either category separately.

Figure 6-9 plots the minimal RMSE among SLG, OBP, and OPS for each year from 1876 through 1999. This is similar to Figure 6-8, except AVG has been eliminated from consideration, and OPS has taken its place. Clearly, SLG and OBP taken together as OPS produce a far-superior model than using either individually. A typical RMSE in twentieth-century baseball is about .15 Runs per Game. This means that using OPS, the number of runs scored by a team per game can be predicted within about .15 Runs per Game for two-thirds of the teams. However, OBP appears to be at least on a par with OPS in predicting runs scored for nineteenth-century teams. (At this point, we can eliminate SLG from consideration before proceeding to the next challenger.)

Total Average (TA)

Another model that has received a lot of attention is *Total Average (TA)*, introduced by sportswriter Thomas Boswell in 1981. TA is a modification of Slugging Percentage. Where SLG is the ratio of total bases to at-bats, TA is the ratio of total bases to total outs. The logic of substituting outs for at-bats is a powerful

Decade	Player	OPS					
1880s	Roger Connor	.883	Pete Browning	.869	Cap Anson	.838
1890s	Dan Brouthers	.942	Ed Delahanty	.917	Sam Thompson	.888
1900s	Honus Wagner	.857	Nap Lajoie	.847	Elmer Flick	.834
1910s	Ty Cobb	.945	Joe Jackson	.940	Tris Speaker	.928
1920s	Babe Ruth	1.163	Rogers Hornsby	1.010	Hack Wilson	.940
1930s	Lou Gehrig	1.080	Jimmie Foxx	1.038	Hank Greenberg	1.017
1940s	Ted Williams	1.115	Joe Dimaggio	.977	Johnny Mize	.959
1950s	Mickey Mantle	.977	Stan Musial	.976	Ralph Kiner	.946
1960s	Willie Mays	.941	Hank Aaron	.928	Frank Robinson	.926
1970s	Dick Allen	.912	Willie Stargell	.889	Reggie Smith	.855
1980s	Mike Schmidt	.908	George Brett	.857	Jack Clark	.854
1990s	Frank Thomas	1.013	Mark McGwire	.981	Barry Bonds	.968

TABLE 6-12   Batters with Highest Career On-Base Plus Slugging from Each Decade

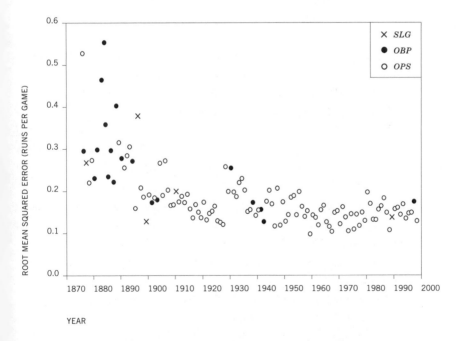

FIGURE 6-9   RMSE of fit to Team Runs per Game since 1876. (Minimum RMSE from SLG, OBP, and OPS = OBP + SLG.)

one. For the most part, the number of outs per game is a constant 27. Thus, it seems natural that the number of runs generated per game should be related to the number of runs per out, which in turn could be related to the number of bases generated per out. This relationship should be tighter than total bases per at-bat since the number of at-bats is more variable over games. Boswell calculated outs by subtracting hits from at-bats and adding Caught Stealing (CS) and Grounded Into Double Plays (GIDP). Like others before and after him, Boswell expanded his model beyond Total Bases (TB) achieved by hitting to include other aspects of offense such as walks (BB), Hit By Pitcher (HBP), and Stolen Bases (SB). The formula for Total Average is as follows:

$$TA = \frac{TB + BB + HBP + SB}{AB - H + CS + GIDP}$$

It should be noted that the concept of using a bases-to-outs ratio to rate offensive performance had been introduced earlier in 1979 by Barry Codell in SABR's *Baseball Research Journal*. Codell's Base-Out Percentage is identical to TA except that the number of sacrifice hits (SH) and sacrifice flies (SF) are added to both the numerator (bases) and denominator (outs). This is consistent with the basic concept, since sacrifice hits and sacrifice flies are both outs that advance runners. (One could argue that CS and GIDP should subtract a base as well as adding an out.) However, TA is the more popular formulation of this concept and the one we will analyze here. As expected, the two formulations provide very similar results. (Because team data on Grounded Into Double Plays was not available, GIDP was assumed to be 0 in the calculation of TA here.)

First, let's look at the Total Average leaders in each decade, as shown in Table 6-13. Rickey Henderson has reappeared at the top of the 1980s players. Tim Raines, an offensive player somewhat like Henderson, now appears in third place in the 1980s. Both Henderson and Raines are productive hitters with some power and tremendous base-stealing abilities. Barry Bonds, another speed demon, has moved into second place among 1990s players. Stealing also allows Jeff Bagwell to slip ahead of Mark McGwire into third place among current players. Clearly, TA's inclusion of SB and CS has given an edge to players whose base stealing is a significant part of their game.

Figure 6-10 shows that OBP, OPS, and TA are equally capable from 1876 to the mid-1890s. Then, OPS dominates into the mid-1930s. Total Average has the edge from then up to the present day. In fact, TA has been the best estimator since 1991.

Decade	Player	TA					
1880s	Roger Connor	.952	Pete Browning	.946	Harry Stovey	.924
1890s	Billy Hamilton	1.191	Dan Brouthers	1.061	Ed Delahanty	1.035
1900s	Honus Wagner	.949	Elmer Flick	.915	Fred Clarke	.883
1910s	Ty Cobb	1.090	Tris Speaker	1.030	Joe Jackson	1.027
1920s	Babe Ruth	1.420	Rogers Hornsby	1.118	Hack Wilson	1.011
1930s	Lou Gehrig	1.248	Jimmie Foxx	1.171	Hank Greenberg	1.133
1940s	Ted Williams	1.374	Joe Dimaggio	1.043	Johnny Mize	1.028
1950s	Mickey Mantle	1.120	Stan Musial	1.067	Ralph Kiner	1.041
1960s	Willie Mays	1.027	Frank Robinson	1.008	Hank Aaron	.984
1970s	Dick Allen	.975	Joe Morgan	.947	Willie Stargell	.914
1980s	Rickey Henderson	1.030	Mike Schmidt	.993	Tim Raines	.931
1990s	Frank Thomas	1.180	Barry Bonds	1.146	Jeff Bagwell	1.103

TABLE 6-13   Players with Highest Career Total Average from Each Decade

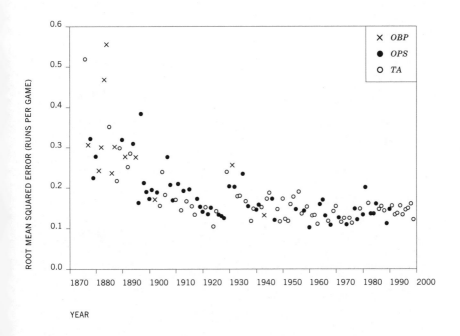

FIGURE 6-10   RMSE of fit to Team Runs per Game since 1876. (Minimum RMSE from OBP, OPS, and TA.)

Batter's Run Average (BRA) and Scoring Index (DX)

TA, then, has a slight edge over OPS. But perhaps a better model than OPS can be created by combining OBP and SLG in a different way. Richard Cramer and Pete Palmer did just that when they *multiplied* OBP and SLG to create *Batter's Run Average* (with the infelicitous acronym *BRA*):

$$BRA = OBP \times SLG$$

The idea here is that scoring runs is the product of getting on-base (OBP) and advancing the runners (SLG). Figure 6-11 plots the minimal RMSE among the OPS, TA, and BRA models. BRA appears to be less effective than TA, but more effective than OPS. So, it seems that multiplying OPS and SLG produces a better model than adding the two values.

Cramer and Palmer were neither the first nor the last researchers to create a model based on this concept. Two of the most notable researchers also adopt this principle in their models. Earnshaw Cook was a metallurgist with a great interest in baseball statistics. His 1964 book *Percentage Baseball* was the first work in baseball statistics to gain the attention of sportswriters and the national

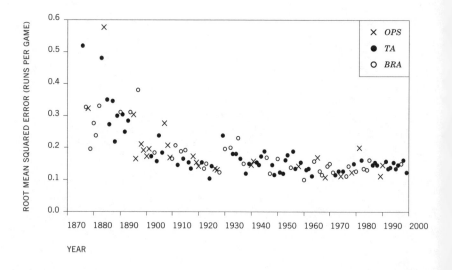

FIGURE 6-11   RMSE of fit to team runs per game since 1876. (Minimum RMSE from OPS, TA, and BRA.)

media. The volume overflows with Cook's ideas, and his enthusiasm for his subject is evident throughout. However, at times, this energy obscures the clarity of his exposition. Perhaps Cook's most lasting contribution was his development of the *Scoring Index (DX)*. His original concept of DX can be expressed this way:

$$DX = \frac{1B + BB + HBP + E}{BFP} \times \frac{TB + SB}{BFP}$$

BFP is the number of times a player came to bat (Batter Faced Pitcher) and E is the number of times the player was safe on an error. DX was developed to be linearly related to runs scored per BFP. He altered the formula in 1971 to this:

$$DX = \frac{H + BB + HBP}{AB + BB + HBP} \times \frac{TB + SB - CS}{AB + BB + HBP}$$

The biggest change in this revised expression is the substitution of Hits (H) for singles (1B) in the first term. The formula is now very similar to BRA, with the inclusion of base-stealing data.

Runs Created (RC)

The other sabermetric heavyweight to adopt this concept was Bill James (who in fact invented the term "sabermetrics" in honor of SABR). The basic concept of James' *Runs Created (RC)* model is as follows:

$$RC = \frac{(H + BB)\,TB}{AB + BB}$$

Since SLG = TB/AB, we see that RC is approximately the same as BRA × AB. So, RC estimates the total number of runs produced while BRA and DX estimate the rate of run production per at-bat or plate appearance. In 1985, James got *really* serious, as evidenced by his technical version of RC:

$$RC = \frac{(H + BB + HBP - CS - GIDP)\,(TB + .26\,(BB - IBB + HBP) + .52\,(SH + SF + SB))}{AB + BB + HBP + SH + SF}$$

The technical RC model (dubbed TECH-1) came with 13 additional versions (TECH-2 through TECH-14) to handle seasons in which some data was not available. Most of the modifications to the original formula are common-sense adjustments (e.g., subtracting runners eliminated by caught stealing and double plays from the on-base term). Unless indicated otherwise, references to the Runs Created model will use the TECH-1 version.

In his *1984 Baseball Abstract,* James indicates that the .26 and .52 multipliers were chosen empirically to improve the fit of Runs Created to total runs scored within each league. In performing this optimization with respect to data, James has moved beyond the realm of intuitive techniques and crossed into data analysis. (Techniques such as these, which involve linear regression, are covered in the next chapter, but since RC is primarily a result of intuition, we will cover it here.)

Since RC predicts total run production, we divide it by the number of games (to obtain RC/G) before fitting it to Runs per Game and comparing it with the other models. Figure 6-12 plots the minimal RMSE values among the TA, BRA, and RC/G models for team run production. RC/G is superior to BRA, as it should be, since it is basically BRA with more data included in the calculation as well as the optimized weights .26 and .52.[5] And RC/G also seems to have an edge over TA.

How does RC/G rate players decade-by-decade? Table 6-14 lists the results of our calculation for individual players, in which we divided RC by an estimate of

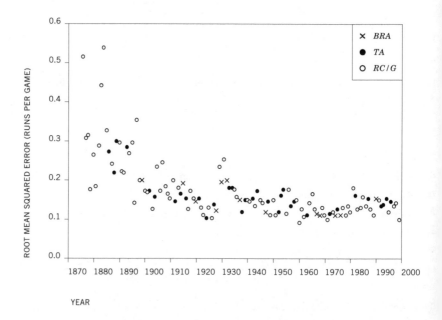

FIGURE 6-12   RMSE of fit to team runs per game since 1876. (Minimum RMSE from TA, BRA and RC/G.)

[5] This raises the issue of whether RC/G provides enough improvement in its prediction to justify the increased complexity of its calculation. This statistical issue is important but beyond the scope of this book.

Decade	Player	RC/G					
1880s	Pete Browning	8.61	Roger Connor	8.41	Cap Anson	7.62
1890s	Billy Hamilton	10.40	Dan Brouthers	9.96	Ed Delahanty	9.50
1900s	Honus Wagner	8.02	Elmer Flick	7.55	Nap Lajoie	7.51
1910s	Ty Cobb	9.77	Joe Jackson	9.26	Tris Speaker	9.05
1920s	Babe Ruth	13.87	Rogers Hornsby	10.56	Hack Wilson	8.75
1930s	Lou Gehrig	11.84	Jimmie Foxx	10.85	Hank Greenberg	10.31
1940s	Ted Williams	13.78	Joe Dimaggio	9.68	Johnny Mize	9.32
1950s	Stan Musial	9.89	Mickey Mantle	9.79	Ralph Kiner	8.93
1960s	Willie Mays	8.62	Frank Robinson	8.39	Hank Aaron	8.30
1970s	Dick Allen	7.99	Willie Stargell	7.42	Joe Morgan	7.00
1980s	Mike Schmidt	7.77	Rickey Henderson	7.49	George Brett	7.14
1990s	Frank Thomas	10.73	Edgar Martinez	9.45	Jeff Bagwell	9.34

TABLE 6-14 Players with Highest Career Runs Created per Game from Each Decade

the number of "games" a player's offensive record represents. This can be done by estimating the total number of outs and dividing by 27. So, Runs Created per Game for an individual player may be calculated as follows:

$$RC / G = \frac{RC}{(AB - H + SH + SF + CS + GIDP) / 27}$$

This value represents the total number of runs produced by a team composed solely of the player analyzed.[6] For example, Stan Musial's RC/G is 9.89. According to the RC model, we expect that a team composed entirely of Stan Musial clones in each of the nine batting slots would score an average of 9.89 Runs per Game. Obviously, Stan the Man was a very great hitter, since teams in this period scored less than half this value, or about 4.5 Runs per Game. This list comes very close to a typical fan's perception of great hitters. We might not have expected the presence of Edgar Martinez or Jeff Bagwell, but in general the list does not have lots of surprises.

---

[6] Actually, the values presented in Table 6-14 somewhat overestimate the actual RC/G since GIDP data were not available.

More Analytic Models

This has not been an easy chapter, and we suspect many of you have spent considerable effort following our arguments. You might even feel like you're out of breath, having stretched a double into a triple. So we will take a moment now to take stock of where we are and how we got here.

Up to this point, the models we have examined were constructed not from any statistical analysis but based on a belief, view, or principle that describes, in a common-sense way, how baseball works. The Batting Average, our starting point, is based on the premise that scoring runs is related to how often a player gets hits. The On-Base Percentage expanded this view to include additional ways of getting on base, primarily through bases on balls. The Slugging Percentage took a somewhat different view and expanded Batting Average by weighting hits in accordance with bases obtained. Another model, Total Average, weighted hits *and* included walks and hit by pitch data, with stolen bases thrown in for good measure; TA also is a ratio of good events (bases) to bad events (outs), as opposed to good events as a percentage or expectation with respect to opportunity (at bats or plate appearances). Two models (On-Base Plus Slugging and Batter's Run Average) combine On-Base Percentage and Slugging Percentage into a single measure through addition and multiplication, respectively. Runs Created expanded on the Batter's Run Average, employing a more detailed accounting of events in which batters get on base and advance runners.

Note that all of these models except Runs Created work completely with integer values (simple counts and weights). They combine these counts to establish impacts of (a) getting on-base and (b) advancing runners relative to the degree of opportunity. For the most part, RC does this as well. But here we see the use of non-integer weightings (.26 and .52) for events that advance runners, but not as effectively as hits advance runners.

Figure 6-13 compares the models with respect to their annual RMSEs averaged over the 46 years from 1954 through 1999. This period was chosen because of the completeness of its data and greater relevance to baseball as it is played today. The standard MLB models (Batting Average, Slugging Percentage, and On-Base Percentage) stand out from the rest as having distinctly worse fits than the newer alternative models. The Runs Created per Game model had the lowest average RMSE (.136). However, the other models (BRA, DX, TA, and OPS) have RMSEs not much greater than RC/G's. Are these differences in RMSE significant?

Table 6-15 compares each pair of models with respect to their RMSEs in each year. For each pair of models, the percentage of years in which the model on the row had a lower RMSE than the model in the column were counted. For example, out of the 46 years of data used, Runs Created per Game had a lower RMSE than Total Average in 70 percent of the data (32 of the 46 years). The question remains whether this difference is significant. Perhaps the two models are equally capable, but chance gave RC/G the edge. If the two models have equal capability to estimate run production, then in each year there is a 50/50 chance that one or the other will have the lower RMSE. Under this equal-capability assumption, there is only a 1.1-percent chance that one of these models will dominate the other with a 32 to 14 edge or greater. This is a very small chance, and so we conclude that in the years 1954–1999, RC/G is a superior model to Total Average. In fact, since it dominates all the other models to an even greater extent, it seems that RC/G provides the best fit to team data of the models considered so far.

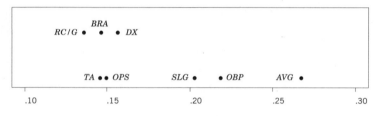

AVERAGE RMSE (RUNS PER GAME)

FIGURE 6-13   Average yearly RMSEs for various models (1954–1999).

Model	RC/G	TA	BRA	OPS	DX	SLG	OBP	AVG
RC/G	·	70%	80%	83%	87%	98%	100%	100%
TA	·	·	57%	61%	72%	93%	96%	100%
BRA	·	·	·	74%	72%	98%	100%	100%
OPS	·	·	·	·	67%	98%	98%	100%
DX	·	·	·	·	·	93%	96%	100%
SLG	·	·	·	·	·	·	61%	96%
OBP	·	·	·	·	·	·	·	89%
AVG	·	·	·	·	·	·	·	·

TABLE 6-15   Percentage of Years That Row Model Had Lower RMSE Than Column Model (1954–1999)

Except for TA vs BRA, TA vs. OPS, and SLG vs. OBP, all differences in model performance are statistically significant in that the probability of getting the result is less than 5 percent.[7] TA, BRA, and OPS are comparable and do very well, considering their simplicity relative to RC/G. Cook's DX does not perform as well as this group, but it is a definite improvement over the MLB standard statistics, which bring up the rear. The question remains, how much further can we improve on these models by applying statistical analysis techniques? The improvement in the Runs Created model gained through the use of optimization gives us some cause for optimism. So far (except for the optimization used in the Runs Created model), we have used statistics merely to *evaluate* models; now we wish to employ statistics to *construct* them.

But before we continue, let's see how our models rated Rocky Colavito and Harvey Kuenn in 1959. Table 6-16 provides ratings for the major models considered in this chapter. Of course Kuenn, as the 1959 American League batting champion, had a higher AVG than Colavito. Colavito's HR power provided him with a slight edge over Kuenn in SLG. But this is the only model in which Colavito dominated. All of the alternatives (OPS, TA, and RC/G) rated Kuenn distinctly higher. One has to conclude that when all aspects of offense are considered, Kuenn gave the greater offensive performance in 1959. Unfortunately, looking forward, 1959 turned out to be a career year for Kuenn. The performances of both Kuenn and Colavito dropped for their new teams in 1960. Still, all of the models rated Kuenn's 1960 performance better than Colavito's. In the subsequent five years, though, Colavito had at least four seasons which topped his 1959 season, while Kuenn, who was traded to a variety of teams, never regained his 1959 form. Ironically, Cleveland made a good trade on the basis of 1959 performance, but Detroit got the better performer in the early sixties.

Player	BA	OBP	SP	OPS	TA	RC/G
Rocky Colavito	.257	.337	.512	.849	.857	6.512
Harvey Kuenn	.353	.402	.501	.903	.923	8.668

TABLE 6-16 Offensive Model Ratings for Rocky Colavito and Harvey Kuenn in 1959

[7] In statistics, this is called a .05 *level of significance*. A level of significance is a quantitative evaluation of the strength of the data when testing a hypothesis. Here, the hypothesis is that the two models being compared have the same capability in predicting team run production. The level of significance specifies that we will accept this hypothesis as being true *unless* the data shows that this hypothesis is extremely unlikely (has probability less than the level of significance). The lower the level of significance, the more proof we are demanding before we reject the hypothesis. All statistical tests are based on this premise, which is similar to our legal tradition's principle that a person (hypothesis) is innocent (true) until proven guilty (false).

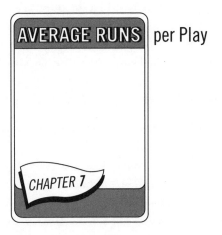

AVERAGE RUNS per Play

CHAPTER 7

Many models developed to evaluate hitting productivity sum the bases attained, then divide that sum by an appropriate measure of opportunity (for example, at-bats). The simplest example is the *Slugging Percentage (SLG)*, which is the total bases attained by hits divided by at-bats:

$$SLG = \frac{1B + (2 \times 2B) + (3 \times 3B) + (4 \times HR)}{AB}$$

One of the best models from Chapter 6, Total Average (TA), was a more complicated example of this form. In this chapter, we will look at how additive-type models like SLG and TA (which sum weighted play frequencies) can be improved by using actual baseball data in their modification. The three data analysis techniques arrive at very similar conclusions (for the most part). In the end, we will have found estimates of the average number of runs that each play event can be expected to produce.

Finding Weights for Plays

Looked at in a more general way, what we have been doing in several different models is to give each event in a game a certain *weight*. In the example of Slugging Percentage, we weight hits, with each type of hit having a weight equal to the number of bases attained: 1 for singles, 2 for doubles, 3 for triples, and 4 for home runs. Each weight is a measure of the impact of the hitting event within the model. So, the SLG model is based on the premise that a home run has 4 times the impact of a single, but only 2 times the impact of a double. Other models have done similar things. Total Average uses the same weights for hits as SLG, and adds walks, hit by pitcher, and stolen bases—each with a weight of 1.

Least Squares Linear Regression (LS LR)

If we generalize the idea of assigning weights to particular play events, and if we don't restrict ourselves to integer weights[1] (1, 2, 3, etc.), we can use a powerful statistical technique to optimize the assignment of weights to each play. In Chapter 6, we used this statistical technique to find the line which minimized the Root Mean Square Error for a model. Without delving into any detail, we were in fact using a statistical technique called *least squares linear regression*. The term itself captures the technique's salient features.. The word "linear" refers to the line which is constructed. The phrase "least squares" reflects the technique's capability to minimize the sum of the squared errors (which forms the basis of RMSE). So far we have used this technique just to gauge how well each model estimates team run production. But the technique is more powerful than we have let on. As many researchers have discovered, least squares linear regression can be used to construct as well as measure models. And not just any models. It constructs the best possible linear model with respect to RMSE.

You may recall that the definition of Total Average is as follows:

$$TA = \frac{TB + BB + HBP + SB}{AB - H + CS + GIDP}$$

Since TB is just total bases obtained from hits, this is really equivalent to the following:

$$TA = \frac{[1B + (2 \times 2B) + (3 \times 3B) + (4 \times HR)] + BB + HBP + SB}{AB - H + CS + GIDP}$$

In reality, all of the play events have weights, so let's put them in explicitly:

$$TA = \frac{[(1 \times 1B) + (2 \times 2B) + (3 \times 3B) + (4 \times HR)] + [(1 \times BB) + (1 \times HBP) + (1 \times SB)]}{AB - H + CS + GIDP}$$

In the previous chapter we found that TA was one of the better models in estimating team run production. It performed well in the sense that over the course of baseball history, the difference between its estimates and the actual run pro-

[1] As James did when developing his Tech-1 version of Runs Created.

duction was better than that of estimates based on standard offensive measures like Batting Average. Still, the question remains whether it is possible to improve on TA. One way to do this is to preserve the basic pattern of summing up weighted numbers of the different events used by TA, but simply change the weights. For example, maybe TA would provide better estimates if the weight for HR was increased from 4 to 5 or reduced from 4 to 3.5 or reduced from 4 to 3.8, or . . . As you can see, the possible choices are infinite. And that's just one weight. There are six others that can also be adjusted. Actually, the great strength of least squares linear regression lies in its ability to *guarantee* finding the weights that produce the best estimate for a given pattern such as TA.

So far, we haven't deviated at all from the Total Average definition. But here we'll make a minor change. The TA denominator (AB − H + CS + GIDP) is the total number of outs, ignoring sacrifice hits and sacrifice flies. Since each team gets approximately the same number of outs per game over the course of a season, we can replace this with G, the number of games.[2] Having made this change, we will now call our model *LSLR* for *Least Squares Linear Regression* model:

$$LSLR = \frac{(1 \times 1B) + (2 \times 2B) + (3 \times 3B) + (4 \times HR) + (1 \times BB) + (1 \times HBP) + (1 \times SB)}{G}$$

Our goal is to find weights for singles, doubles, triples, etc., that improve on the TA estimates. So, we have to generalize the weights to arbitrary values to which we have given the names w_{event}. For example, currently, w_{HR} has the value 4, but we will find a new value for w_{HR} which minimizes the error in estimates. In addition, since there is little (if any) difference between walks and hit by pitcher we will use the same weight w_{BB} for both events. So, for now, until we find those values, we now have:

$$LSLR = \frac{(W_{1B} \times 1B) + (W_{2B} \times 2B) + (W_{3B} \times 3B) + (W_{HR} \times HR) + [W_{BB} \times (BB + HBP)] + (W_{SB} \times SB)}{G}$$

Finally, we will divide each event by the number of games:

$$LSLR = \left(W_{1B} \times \frac{1B}{G}\right) + \left(W_{2B} \times \frac{2B}{G}\right) + \left(W_{3B} \times \frac{3B}{G}\right) + \left(W_{HR} \times \frac{HR}{G}\right) + \left(W_{BB} \times \frac{BB + HBP}{G}\right) +$$
$$\left(W_{SB} \times \frac{SB}{G}\right)$$

---

[2] Since theoretically each team has 27 outs in each game, the number of games G is approximately equal to (AB − H + CS + GIDP)/27. For our purposes, the division by 27 is not needed; 27 is a scaling factor which is constant for all teams.

Weight	w_{SB}	w_{BB}	w_{1B}	w_{2B}	w_{3B}	w_{HR}	RMSE
TA	1	1	1	2	3	4	.159
LSLR	.16	.36	.52	.67	1.18	1.50	.142

TABLE 7-1   Weights in Total Average and in Linear Least Square Regression Model Fit to 1954–1999 Team Runs per Game

This form works best for least squares linear regression. Notice that we now have the model expressed simply as a sum of weighted quantities; here the quantities are frequencies of different events per game.

We will not go into any details about how the best weights are calculated. Descriptions of regression algorithms can be found in a standard statistics textbook. The results of the calculation are shown in Table 7-1 with respect to team data from the 1954–1999 seasons. The weights found for LSLR are very different in scale from the TA weights. This is because the regression techniques scale the weights using the same scale as the values being estimated. So, the LSLR weights are in terms of runs. For example, the regression estimates that each triple is worth, on average, 1.18 runs. When comparing the TA and LSLR weights, the important thing to focus on is the *relative* value of the weights within each model.

In the Total Average model, stolen bases, walks, hit by pitcher, and singles all have the same weights. The LSLR model finds a big difference in the values of these events. Stolen bases have a weight less than half that of walks and hit by pitcher, and less than a third that of singles. We may quibble about the exact values of each of these events, but it seems reasonable that singles should have a greater value than walks and hit by pitcher, which in turn should have a greater value than stolen bases. Singles, walks, and hit by pitcher all put the batter on first base, but a single usually advances all runners, while walks and hit by pitcher only advance runners who are forced. Moreover, singles can advance runners two bases, while walks and hit by pitcher are limited to a one-base advance at most. Comparisons between walks or hit by pitcher and stolen bases are less clear-cut. The argument rests on the relative merits of getting on base as opposed to advancing while on base. In most situations, getting on base produces greater run potential than a single runner advancing one base. Besides, walks and HBPs often advance runners in addition to creating another base runner with the opportunity to score.

With respect to hits, LSLR places less weight on extra-base hits than TA does. The LSLR weight for home runs is less than 3 times that of singles (compared to

4 times in TA). The LSLR weight for triples is approximately 2 times that of singles (compared to 3 times in TA), and the LSLR weight for doubles is roughly 1.3 times that of singles (compared to 2 times in TA). It appears that LSLR finds additional value in getting on base beyond the number of bases attained. This is similar to the position taken by OPS, which adds On-Base Percentage to Slugging Percentage, thereby creating an overall effect similar to that found in the LSLR weights for hits.

On the whole, the LSLR weights make sense. But how well do they estimate team run production? As expected, the RMSE for LSLR is less than that of TA. After all, regression techniques are guaranteed to find the weights that minimize RMSE. Remember that TA and LSLR are virtually identical in form; their only difference is in the values given to the weights for the various events. What makes linear regression so powerful is that it requires no knowledge at all about baseball. In order to create Total Average, Thomas Boswell utilized his insight into baseball as an experienced sportswriter to distill what he thought were the essential elements of run productivity. And his model performed quite well. But linear regression was able to construct an even better model without the researcher having any understanding at all about baseball. We could give the team baseball data to a Greek statistician who has never seen a baseball game, who doesn't know what a strike, single, or out is, and that statistician would be able to develop the weights for this model as capably as the most knowledgeable sabermetrician can.[3]

The property of the LSLR model producing the lowest possible RMSE is guaranteed only for the set of data with which it was derived. This means that the LSLR model in Table 7-1 is the best linear model *only for the 1954 through 1999 seasons*. Employing the same graphical technique from Chapter 6, Figure 7-1 plots the lower RMSE (from TA or LSLR) when estimating team runs in each separate season. In the period 1954–1999, we find that LSLR had a lower RMSE than TA in 35 of the 46 seasons. So, although LSLR was better overall in that period (as guaranteed by the least squares fit), there were individual years (as recently as 1996) where TA had the better fit.

More important is finding out how well LSLR estimates run productivity in the years *before* 1954. This is important because it provides a way to *validate* the LSLR weights independently—with a set of data different from the data used to develop the model. As it turns out, LSLR has a lower RMSE than TA in 57 out of

---

[3] Of course, interpreting the results of the regression would be quite challenging for a Greek statistician (or any statistician) who was not familiar with baseball.

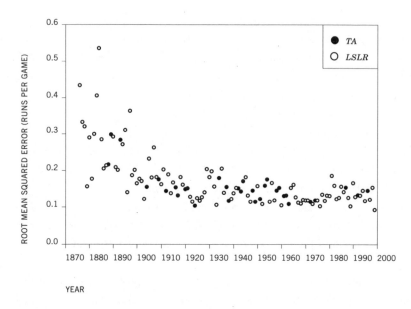

FIGURE 7-1   RMSE of fit to team runs per game since 1876 (Minimum RMSE of TA and LSLR).

the 78 seasons from 1876–1953. Its fit compared to that of TA in this earlier span of seasons is almost as good as it was in 1954–1999 (better in 73 percent versus 76 percent of the years). Given the more reasonable nature of its weights and the improved fit over all of baseball history, we must conclude that LSLR is superior to TA as a model for team run production.

We can make an even stronger statement about LSLR. It is not only better than Total Average, but it is better *than any other possible* additive model that uses the same data in a linear fashion (by summing the weighted frequencies of plays). In fact, LSLR is superior to the whole family of additive models, including:

- Batting Average and On-Base Percentage, where each on-base event has the same weight;

- Slugging Percentage, where each hit is weighted by the number of bases; and

- OPS, where each walk and hit by pitch has a weight near 1, and each hit has a weight near 1 plus the number of bases attained (e.g., a double has a weight near 1 + 2 = 3).

We can't yet say, however, that LSLR as a model provides a better fit than all weighted models. The *Batter's Run Average (DX)* and *Runs Created (RC)* models, which are based on *multiplying* OBP and SLG, do not belong to the additive class. Let's look first at the basic Runs Created model:

$$RC = \frac{(H + BB)\ TB}{AB + BB}$$

If we expand H and TB into the individual play counts, we have:

$$RC = \frac{(1B + 2B + 3B + HR + BB)\ [1B + (2 \times 2B) + (3 \times 3B) + (4 \times HR)]}{AB + BB}$$

This multiplicative form is different from that of LSLR, so it is possible that regression applied to this version of the RC model could have a better RMSE than .142 runs per game. As it turns out, for the period from 1954–1999, *Runs Created per Game (RC/G)*,[4] the best model we have examined in this class, has an RMSE of .146 runs per game. Figure 7-2 plots the RMSE for the model with the lower RMSE when estimating team runs in each separate season. (LSLR has the lower RMSE in 27 out of the most recent 46 seasons.) Using only the results from 1876–1953 as a test of model superiority, LSLR has a lower RMSE in 43 of the first 78 seasons in baseball history. This is not strong enough proof that LSLR is superior to RC/G, but the result does indicate that LSLR is at least as good as RC/G in estimating team run productivity.

These results are both good and bad news. Good in the sense that we have found the best model, but bad in the sense that we now know that we cannot do any better than an RMSE of about 0.14 runs per game using an additive type model. Remember that two-thirds of the observed team runs per game fall within one RMSE of the values predicted by the LSLR model. So, out of all the team predictions in the last 46 years of MLB history, LSLR is correct within .14 runs per game for two-thirds of all estimates. Looking back at team runs per game in the 1998 season, we see that changing a team's run productivity by .14 runs per game could move the team up or down as many as 6 places in Table 6-4. This accuracy cannot be topped by any other linear model with this data. While it is possible that a model of some other form could produce better estimates, the

---

[4] Since RC (as well as the BRA and DX models) can be described as a weighted sum of cross product terms, it is possible to use linear regression techniques to find better weights for hits, walks, and stolen bases in those models as well. Using the RC Tech-1 model form, we found that the 1954–1999 RMSE could be reduced to about .140 runs per game using better weights. Interestingly, the weights for walks, hit by pitcher, and stolen bases were not far different from the .26 and .52 values in the RC Tech-1 model. This is an independent verification of the effectiveness of James' empirical development of these weights.

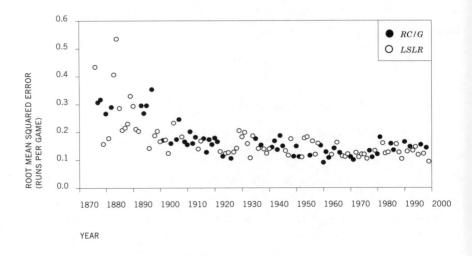

FIGURE 7-2   RMSE of fit to team runs per game since 1876 (Minimum RMSE of
LSLR and RC/G).

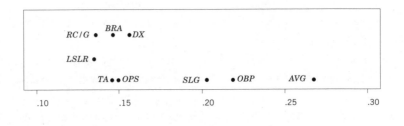

FIGURE 7-3   Average yearly RMSEs for various models (1954–1999).

ones we have examined (BRA, DX, and RC) do not provide any improvement. If
we add LSLR to the dotplot in Figure 6-13, we find (not surprisingly) that LSLR
has the lowest average RMSE of all models considered so far. (See Figure 7-3.)

Adding Caught Stealing to the LSLR Model

There is one way that we can improve the LSLR model's fit: by using additional information. One piece of information that we have not used is the number of times runners have been *caught stealing (CS)*. Since we have this data

Weight (Runs)	w_{CS}	w_{SB}	w_{BB}	w_{1B}	w_{2B}	w_{3B}	w_{HR}	RMSE
With CS	−.11	.19	.35	.52	.66	1.17	1.49	.1421
Without CS	•	.16	.36	.52	.67	1.18	1.50	.1423

TABLE 7-2   Least Squares Linear Regression Model (With and Without Caught Stealing) Fit to 1954–1999 Team Runs per Game

for 1954–1999, it is easy to modify the LSLR model to use this information. All we have to do is add the appropriate CS per game rate multiplied by a new weight, w_{CS}:

$$LSLR = \left(W_{1B} \times \frac{1B}{G}\right) + \left(W_{2B} \times \frac{2B}{G}\right) + \left(W_{3B} \times \frac{3B}{G}\right) + \left(W_{HR} \times \frac{HR}{G}\right) + \left(W_{BB} \times \frac{BB + HBP}{G}\right) +$$

$$\left(W_{SB} \times \frac{SB}{G}\right) + \left(W_{CS} \times \frac{CS}{G}\right)$$

Using regression techniques on data from 1954–1999, we obtain the results in Table 7-2 for LSLR with CS. (For comparison, we have also included the earlier results for LSLR without CS.) Since being caught stealing has a detrimental effect on scoring runs, unlike the other weights, CS has a *negative* weight. And in absolute terms, it has the smallest of all weights. We find that including CS has indeed decreased RMSE, it has done so by a disappointingly small amount, from .1423 to .1421. Why is this so?

One way to investigate this almost negligible impact is to delve deeper into the relationships among the play event quantities—that is, how the frequency of one event type is related to the frequency of another. Figure 7-4 plots caught stealing per game (CS/G) versus stolen bases per game (SB/G) for each team in the years 1954–1999. An increasing trend is evident in the graph. The five teams with the highest stolen-base rate per game are noted in the graph. Three of these teams (the 1976 Oakland A's, the 1977 Pittsburgh Pirates, and the 1992 Milwaukee Brewers) are also among the teams with the highest rate of being caught stealing. This should not be surprising to most baseball fans. Teams that are fast try to take advantage of their skills and attempt to steal more bases; so the frequency of stolen bases is high, but the number of times they get caught stealing is also high. Slower teams do not steal as often, so their number of successes and number of failures are both low.

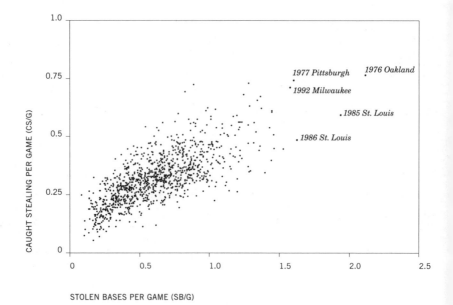

FIGURE 7-4   Team CS/G versus SB/G (caught stealing per game versus stolen bases per game) from 1954–1999.

Such strong correlations make the linear regression weights difficult to interpret. Sometimes the factors are so closely correlated with each other that it is impossible to separate the effects of one from the other. Imagine if the correlation between two factors was even stronger, so strong that the points formed a shape closer to a line. In this case, it would be impossible to determine which factor caused the effect being estimated. Laboratories that conduct research requiring statistical analysis strive to design their experiments to eliminate dependencies between the factors being analyzed. And they often have the luxury of controlling their environment in order to preserve the independence of the factors.[5] However, when data is recorded in an uncontrolled environment (such as a season of baseball games), it is rare that the quantities collected are completely independent of one another.

[5] For example, suppose a new drug is being tested for its effectiveness. If we gave the new drug only to male subjects and a placebo only to female subjects, the two factors, medication and gender, would be completely dependent. Whatever differences we find in subject response, we would not be able to tell whether it was from the new drug or the difference in gender. So, to preserve independence of these factors, we can set up the experiment so that half of each gender gets the new drug and half the placebo. We have the capability to control the factors so that the effects of gender and medication can be separated.

The addition of caught stealing in the LSLR model reduces RMSE very little because most of its information at a team level is already captured by the number of stolen bases. When CS is not included in the LSLR model, we see that the SB weight drops from .19 to .16. Basically, since SB already carries most of the information about caught stealing, it also assumes the negative effects of caught stealing once CS is removed from the model. So we see that when two quantities used in the model are strongly related, the inclusion of one affects the weight of the other. When both are present—as SB and CS were in our original regression model—the two actually compete with each other for dominance of the total weight that they actually share. In conclusion, since the addition of CS provided little improvement in fit, and since it is closely related to SB, which is already in the LSLR model, there is no reason to add it to the LSLR model. But we now understand that the .16 weight for stolen bases encompasses the effects from unsuccessful steal attempts as well as from successful ones.[6]

Adding Sacrifice Flies to the LSLR Model

Maybe we'll have more success with a different piece of information. Let's try *sacrifice flies (SF)*. Just as we did for caught stealing, we just add the appropriate SF per game rate multiplied by a new weight w_{SF}:

$$LSLR = \left(W_{1B} \times \frac{1B}{G} \right) + \left(W_{2B} \times \frac{2B}{G} \right) + \left(W_{3B} \times \frac{3B}{G} \right) + \left(W_{HR} \times \frac{HR}{G} \right) + \left(W_{BB} \times \frac{BB + HBP}{G} \right)$$
$$+ \left(W_{SB} \times \frac{SB}{G} \right) + \left(W_{SF} \times \frac{SF}{G} \right)$$

Table 7-3 shows the weights and RMSE with sacrifice flies included in the regression model. This time we do see some larger changes. RMSE has dropped from .142 to .138. The SF weight itself is large, greater than all plays except triples and home runs. In fact, the weights for all play events except home runs have decreased with the addition of sacrifice flies to the model.

---

[6] You may have noticed slight changes in the weights of other play frequencies when caught stealing was added. This is because CS and the other play events are interrelated. We examined the relationship between stolen bases and caught stealing because it was the most extreme, but there are other quantities which are correlated with one another. The frequency of home runs in particular has strong correlations with other play frequencies. This is especially true of doubles and walks/HBPs. Apparently, teams with more home runs also tend to have more doubles and also reach first on walks and hit batsmen more frequently. However, none is as strongly related as CS to SB, so the effects of these other relationships are relatively small.

Weight (Runs)	w_{SB}	w_{1B}	w_{2B}	w_{SF}	w_{3B}	w_{HR}	w_{BB}	RMSE
Without SF	.16	.52	.67	•	1.18	1.50	.36	.1423
With SF	.14	.49	.61	.73	1.14	1.50	.33	.1381

TABLE 7-3   Least Squares Linear Regression Model (With and Without Sacrifice Flies) Fit to 1954–1999 Team Runs per Game

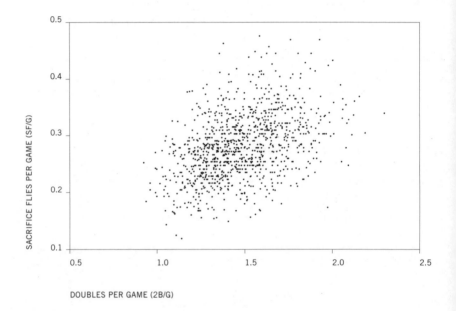

DOUBLES PER GAME (2B/G)

FIGURE 7-5   Team SF/G versus 2B/G (sacrifice flies per game versus doubles per game) from 1954–1999.

The number of sacrifice flies per game is correlated with other play events. The strongest correlation is between sacrifice flies and doubles. Figure 7-5 plots sacrifice flies per game versus doubles per game for teams from 1954–1999. There does appear to be a slight upward trend. Still, none of the leading teams in doubles is also a leader in sacrifice flies, and similarly none of the leading teams in sacrifice flies is also a leader in doubles. The cloud of points seems more shapeless than the pattern seen in CS/G versus SB/G. So, SF/G is not correlated with any other play event rate as strongly as CS/G was with SB/G.

While RMSE has dropped much more than it did with CS, the SF weight seems inordinately high. It is difficult to believe that a play in which the batter is out could be more valuable than a walk, much less a double. This leads us to believe that there is some added information carried by SF that is not contained in the other play events. In regression, the quantities used in the formula may possess more meaning than the analyst originally surmises, and this is why some statisticians refer to these quantities as "carriers." These quantities may carry more information than their literal name implies. For example, perhaps the number of SFs embodies some unknown intangible quality of the team beyond the tendency to hit sacrifice flies.

The definition of sacrifice fly could give us some perspective. Recall that a sacrifice fly *always* drives in a run. This is part of its definition. And it is not surprising that a play which *always* results in a run would be highly correlated with run productivity. But is there some hidden meaning or interpretation of SF data? One question, for example, is whether the SF weight captures the value of the *situation* in which the sacrifice fly tends to occur (a runner on third base with fewer than two outs). Is it possible that this situation, even more than the play itself, ties SF closely to run production?

So we have arrived at a quandary. It is possible to reduce the error in our estimate, but this improvement is obtained by using a weight for sacrifice flies that is suspect. Perhaps further enlightenment will come from taking a more detailed view of individual plays and their influence in scoring runs.

The Lindsey-Palmer Models

Sabermetrics has many notable contributors; Bill James and Pete Palmer are perhaps the best known. And then there is George Lindsey, a Canadian defense consultant who has a great love for baseball. Like many of us who pursued research in this area much later, he was dissatisfied with the state of baseball statistics in the late 1950s. He saw no reason why the quantitative techniques he applied in his day-to-day job could not be used to gain a better understanding of the game. His research papers on baseball, published in the early 1960s, were among the first to appear in scientific journals.

George Lindsey's Analysis

Lindsey's research focused on run production, its effect on winning the game, and the use of this knowledge to determine the effectiveness of various strategies (bunting and stealing, for example). In an age when baseball data was not nearly as available as it is today (remember the first modern baseball encyclo-

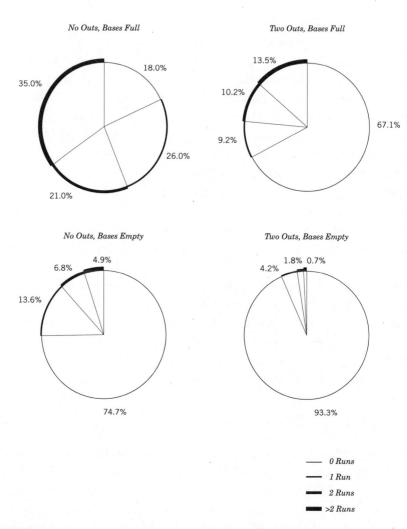

FIGURE 7-6   Probability of scoring runs in selected extreme situations (Lindsey, 1963).

pedia was not published until 1969), Lindsey employed the services of his father to gather play-by-play data for 27,027 situations in 373 games broadcast during the 1959 and 1960 seasons.

His analysis of this play-by-play data produced probability distributions of runs scored in all base-out situations. (One man on, no out; one man on, two out; and so forth through to bases loaded, two out.) Figure 7-6 shows some of his results for the most extreme situations. We can see that the least variable situation is represented by the pie chart in the lower right of the figure—two outs and no runners on base—where the probability of scoring any runs is less than 7 percent. At the other extreme (no outs and bases loaded), the team at bat has an 82 percent chance of scoring runs. While this probability is high, it may not be as high as we expect; scoring runs even in this most advantageous of all situations is not a sure thing. Still, there is twice as great a chance that a big inning (three or more runs) will result than that the team will not score at all. We also compare two other extreme and opposite situations: bases empty with no outs versus bases loaded with two outs. In this pairing, we find that each situation has close to the same probability that no runs will score (75 percent versus 67 percent). But bases loaded with two outs is the more volatile situation, primarily because the chance of scoring three or more runs is almost three times greater than in the bases empty with no outs situation.

These results are interesting by themselves, but Lindsey was only getting started. Table 7-4 shows the 4 probability distributions from Figure 7-6 as well as all of the other 20 base-out situations. Taken together, the two columns "Bases Occupied" and "Outs" identify all 24 possible base-out situations within an inning. The "% of Situations" column gives some feeling for how often each situation occurs. The most common situation is none on and no outs; this is expected, since every inning starts this way. In general, the situations become increasingly rare as the number of base runners, and how far they have advanced, increases.

The middle columns present the probabilities of scoring different numbers of runs in each situation. The column on the far right gives the expected number of runs scored in each situation. These values were calculated by Lindsey from the complete set of run-scoring probabilities, which are presented here for only the most common run totals.[7] As expected, bases loaded with no outs has the greatest expected value (2.22 runs) while no base runners with two outs has the lowest expected value (.102 runs). Further scrutiny of the table shows that a greater

---

[7] Lindsey calculated this by multiplying the runs scored by the probability that they would be scored and then summing these products for all possible runs scored from 0 to ∞. Of course, there is a practical limit to the number of runs scored in an inning, the record being 18, Chicago White Stockings (NL) (vs. Detroit Wolverines), seventh inning, September 6, 1883 (*The Book of Baseball Records*, Seymour Siwoff [ed.], New York: Elias Sports Bureau, 1999). For example, a team is expected on average to score .102 runs when they have no base runners and two outs. This value was calculated as follows:

$$.933 \times 0 \text{ runs} + .042 \times 1 \text{ run} + .018 \times 2 \text{ runs} + .007 \times 3 \text{ runs} =$$
$$.042 \text{ runs} + .036 \text{ runs} + .021 \text{ runs} = .099 \text{ runs}$$

The reason for this slight underestimation of Lindsey's value of .102 runs is that we assumed that the number of runs associated with the probability of scoring 2 or more runs was exactly 3. Lindsey did not publish the separate probabilities of scoring 3 runs, 4 runs, 5 runs, etc., which would have allowed us to reproduce his value exactly.

SITUATION			PROBABILITY OF SCORING RUNS				EXPECTATION
Bases Occupied	Outs	% of Situations	0 Runs	1 Run	2 Runs	>2 Runs	Runs
None	0	24.3%	.747	.136	.068	.049	.461
	1	17.3%	.855	.085	.039	.021	.243
	2	13.7%	.933	.042	.018	.007	.102
1	0	6.4%	.604	.166	.127	.103	.813
	1	7.6%	.734	.124	.092	.050	.498
	2	7.8%	.886	.045	.048	.021	.219
2	0	1.1%	.381	.344	.129	.146	1.194
	1	2.4%	.610	.224	.104	.062	.671
	2	2.9%	.788	.158	.038	.016	.297
3	0	0.2%	.120	.640	.110	.130	1.390
	1	0.7%	.307	.529	.104	.060	.980
	2	1.2%	.738	.208	.030	.024	.355
1,2	0	1.4%	.395	.220	.131	.254	1.471
	1	2.6%	.571	.163	.119	.147	.939
	2	3.3%	.791	.100	.061	.048	.403
1,3	0	0.4%	.130	.410	.180	.280	1.940
	1	1.1%	.367	.400	.105	.128	1.115
	2	1.6%	.717	.167	.045	.071	.532
2,3	0	0.3%	.180	.250	.260	.310	1.960
	1	0.7%	.270	.240	.280	.210	1.560
	2	0.8%	.668	.095	.170	.067	.687
Full	0	0.3%	.180	.260	.210	.350	2.220
	1	0.8%	.303	.242	.172	.283	1.642
	2	1.0%	.671	.092	.102	.135	.823

TABLE 7-4   Distribution of Runs Scored in Remainder of Inning (Lindsey, 1963)

number of base runners does not always compensate for an increase in outs. Only 1.64 runs are expected to be scored with the bases loaded and one out, while no outs with runners on first and third or second and third are expected to produce 1.94 and 1.96 runs, respectively. The possibility of a double play ending the inning with no runs scored is responsible for much of this effect.

INITIAL STATE				FINAL STATE				CHANGE
Bases Occupied	Outs	% of Situations	Expected Runs	Bases Occupied	Outs	Expected Runs	Runs	Total Runs	Runs
None	0	24.3%	.461	None	0	.461	1	1.461	1
	1	17.3%	.243	None	1	.243	1	1.243	1
	2	13.7%	.102	None	2	.102	1	1.102	1
1	0	6.4%	.813	None	0	.461	2	2.461	1.648
	1	7.6%	.498	None	1	.243	2	2.243	1.745
	2	7.8%	.219	None	2	.102	2	2.102	1.883
2	0	1.1%	1.194	None	0	.461	2	2.461	1.267
	1	2.4%	.671	None	1	.243	2	2.243	1.572
	2	2.9%	.297	None	2	.102	2	2.102	1.805
3	0	0.2%	1.390	None	0	.461	2	2.461	1.071
	1	0.7%	.980	None	1	.243	2	2.243	1.263
	2	1.2%	.355	None	2	.102	2	2.102	1.747
1,2	0	1.4%	1.471	None	0	.461	3	3.461	1.990
	1	2.6%	.939	None	1	.243	3	3.243	2.304
	2	3.3%	.403	None	2	.102	3	3.102	2.699
1,3	0	0.4%	1.940	None	0	.461	3	3.461	1.521
	1	1.1%	1.115	None	1	.243	3	3.243	2.128
	2	1.6%	.532	None	2	.102	3	3.102	2.570
2,3	0	0.3%	1.960	None	0	.461	3	3.461	1.501
	1	0.7%	1.560	None	1	.243	3	3.243	1.683
	2	0.8%	.687	None	2	.102	3	3.102	2.415
Full	0	0.3%	2.220	None	0	.461	4	4.461	2.241
	1	0.8%	1.642	None	1	.243	4	4.243	2.601
	2	1.0%	.823	None	2	.102	4	4.102	3.279

TABLE 7-5   The Run Values of Home Runs

Having set up the data as shown in Table 7-4, Lindsey realized that he could use it to estimate the value of each hit in terms of runs. The easiest hit to analyze is the home run, as shown in Table 7-5. At first glance, this table is somewhat daunting, but it is more easily understood once broken down into its three major components: (1) the *Initial State*, (2) the *Final State*, and (3) the *Change in State*.

1. *The Initial State.* The first four columns are carried over from Table 7-4. They describe the initial state when the batter comes to the plate (identified by the bases occupied and the number of outs), the percentage of time that the state occurs, and the average number of runs scored in the inning after a plate appearance under these circumstances.

2. *The Final State.* The next five columns describe the final state, the game situation after the batter's plate appearance. Since we are analyzing the effect of a home run, the outs do not change from the initial state, and the bases are empty in all cases. The third final state column presents the expected number of runs from the final base-out state. These values were found by looking up Expected Runs in Table 7-4 for the base-out situation that exists in the final state. Since bases are always empty after a home run, this has to be one of three values: .461, .243, or .102 runs, depending on the number of outs—0, 1, or 2, respectively.

   The fourth column is the number of runners who scored on this play. For a home run, this is just the number of base runners in the initial state plus 1.

   So, we have two run components: the number of runs that scored on the play and the expected number that may still score in the future, based on the number of outs and the runners left on base after the play. The total value of the final state in terms of runs (displayed in the fifth final state column) is the sum of these two components.

3. *The Change in State.* The third part of Table 7-5 is the change between the two states. Since we are performing this analysis in terms of runs, we interpret this change to mean the difference in run value between the initial state and the final state. The run value of the initial state is the expected number of runs given in the fourth initial state column. The run value of the final state is the total number of runs given in the fifth final state column. The change in runs is calculated as follows:

$$\text{Change} = \text{total runs (final state)} - \text{expected runs (initial state)}$$

This change is the run value of the play (HR in this case) in the particular situation defined by the initial state.

Let's see how we can apply Tables 7-4 and 7-5 to one of the greatest moments in baseball history (the #24 all-time moment as picked by The Sporting News in 1999). In 1986, the California Angels were one out away from their first trip to the World Series. They led the Boston Red Sox 5–4 in the top of the ninth inning, and Dave Henderson was Boston's last chance. The Bosox had a runner on first and two outs. According to Table 7-4, the Red Sox had less than a 5-percent chance of tying the game and about a 7-percent chance of going ahead. Overall, the Red Sox could only be expected to score an average of .219 runs in this situation.

However, Henderson connected for a two-run homer and gave the Red Sox the lead. The final base-out state was two outs with no runners on base. Since the Red Sox still had another out, they had a chance to score more runs. According to Lindsey's data, they could only expect to score .102 more runs on average in the remainder of the inning. Since they already scored two runs from the home run, the expected run value of this final state is $2 + .102 = 2.102$. Subtracting the expected run value of the initial state from that of the final state, we obtain:

$$\text{Change} = \text{total runs (final state)} - \text{expected runs (initial state)}$$
$$= 2.102 - .219 = 1.883$$

To put it another way, before Henderson's HR, the Red Sox had 4 runs with the expectation of scoring .219 more runs. After the HR, the Red Sox had 6 runs with the expectation of scoring .102 more runs. So, the value of the HR in terms of runs is $6.102 - 4.219 = 1.883$ runs.[8] Although Henderson's HR produced two RBIs, its run value was actually less than 2. This is because the initial state with a runner on first base had a run value (.219 runs) greater than the final state (.102 runs), in which the bases were empty after the HR. In fact, looking down the Change in Runs column of Table 7-5, we see that this is true for all HRs. According to Lindsey's model, the RBI statistic overstates the true value of any home run except one hit with the bases empty. In essence, a home run converts all potential runs into actual runs, leaving the bases depleted of all run potential.

Another interesting observation in scanning down the Change in Runs column of Table 7-5 is that the value of HRs increases with the number of outs. For example, Henderson's two-out HR with a runner on first base had greater value

---

[8] The Red Sox did not score more runs that inning, but they did go on to win the game and the AL Championship Series only to be thwarted themselves by ill-luck at the hands of the New York Mets (or the hands of Bill Buckner) in the 1986 World Series.

(1.883 runs) than if he had hit it with one out (1.745 runs) or no outs (1.648 runs). This makes sense. As outs increase, the opportunities for putting runners on-base and advancing them to score decreases. Lindsey's model provides quantitative support for our intuitive feel for the game.

To proceed with the analysis, we assume that HR frequency is independent of the situation; that is, a home run is equally likely to occur in any base-out situation. This is a big assumption, but one that is standard in most baseball models. Under this assumption, we can construct a distribution of HR values based on the initial state and Change in Runs column of Table 7-5.

We have done just this in Table 7-6, where the rows representing the initial states are sorted by run value ("Change in Runs"). Each row also includes the number of occurrences observed for each kind of situation ("# of Situations") followed by its percentage ("% of Situations") as compared to the total number of situations observed. We see that the greatest value of a home run is 3.279 runs (bases loaded, two outs), less than the 4 RBIs credited to a grand slam. This maximum value is found in only 283/27,027 = 1 percent of all HRs. The smallest value is 1 run (bases empty with 0, 1, or 2 outs). This value is also the *mode* of the distribution (the most common value). It occurs in 14,935 out of the 27,027 situations used to create Table 7-5; so, 55 percent (a majority) of HRs have a value of 1 run.

Note that a home run with no outs and a runner on third does not have much greater value than one with the bases empty. This makes sense, since it is expected that this runner should be able to score anyway, with 3 outs available to the offense; the HR contributes a minimal amount to scoring this runner. In fact, the lower HR values are dominated by situations with 0 or 1 out and runners in scoring position. The higher HR values are dominated by situations with multiple runners and two outs; these are high-risk situations where there is much to gain but little opportunity to do so.

The "Cum % of Situations" column in Table 7-6 tracks the percent of situations with HR values less than or equal to that of the current row. A visual perspective of the HR value distribution can be obtained by plotting this column as the cumulative probability of HR value. Such a plot (Figure 7-7) shows us how the probability grows as larger and larger HR values are accumulated, starting from 0 runs. The line in the plot indicates the probability that a HR has a value less than or equal to a given number of runs (indicated on the x-axis). We see that the probability is 0 until 1 run is reached, since it is impossible for any home run to be worth less than 1 run. At 1 run, the cumulative distribution jumps up to 55 percent; as mentioned earlier, there is a 55-percent chance that a

Bases Occupied	Outs	# of Situations	% of Situations	Cum. % of Situations	Change in Runs
None	0	6561	24.3%	24%	1
None	1	4664	17.3%	42%	1
None	2	3710	13.7%	55%	1
3	0	67	0.2%	56%	1.071
3	1	202	0.7%	56%	1.263
2	0	294	1.1%	57%	1.267
2, 3	0	73	0.3%	58%	1.501
1, 3	0	119	0.4%	58%	1.521
2	1	657	2.4%	60%	1.572
1	0	1728	6.4%	67%	1.648
2, 3	1	176	0.7%	68%	1.683
1	1	2063	7.6%	75%	1.745
3	2	327	1.2%	76%	1.747
2	2	779	2.9%	79%	1.805
1	2	2119	7.8%	87%	1.883
1, 2	0	367	1.4%	88%	1.990
1, 3	1	305	1.1%	90%	2.128
Full	0	92	0.3%	90%	2.241
1, 2	1	700	2.6%	92%	2.304
2, 3	2	211	0.8%	93%	2.415
1, 3	2	419	1.6%	95%	2.570
Full	1	215	0.8%	96%	2.601
1, 2	2	896	3.3%	99%	2.699
Full	2	283	1.0%	100%	3.279

TABLE 7-6   Distribution of Run Values for Home Runs

home run is worth exactly 1 run. The plot keeps increasing as HRs with greater and greater value are included. It does not rise steeply at first. There is still only about a 57-percent chance for a HR to have value less than 1.5 runs, not much of an increase over the 55 percent of HRs with value less than or equal to 1 run. So, there is only a 57 percent – 55 percent = 2-percent chance that a HR is worth between 1 and 1.5 runs. Between 1.5 and 2 runs, the plot rises quickly; there is about an 88-percent chance that a HR has a value less than or equal to 2 runs. This shows us that there is a good chance (88 percent – 57 percent = 31 percent) that a HR is worth between 1.5 and 2 runs. From here, the graph rises steadily until it

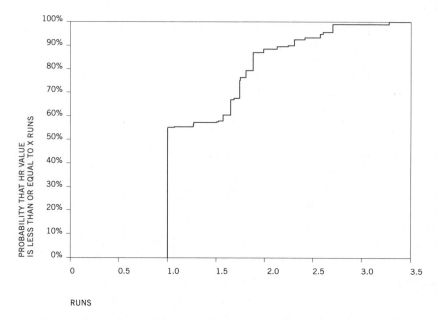

RUNS

FIGURE 7-7   Cumulative distribution of run values for home runs.

reaches 100 percent at the maximum HR value of 3.279; since this is the maximum, 100 percent of HR values are less than or equal to it and 12 percent (100 percent − 88 percent) are worth between 2 and 3.3 runs.

Using Table 7-6 (or Table 7-5), we can calculate the average value of a home run. This is done by simply multiplying its value in each situation by the percentage of times the situation occurs, then summing over all situations:

Average HR Value = $(.243 \times 1) + (.173 \times 1) + (.137 \times 1) + (.002 \times 1.071) + \ldots$
$$+ (.008 \times 2.601) + (.033 \times 2.699) + (.01 \times 3.279) = 1.42 \text{ runs}$$

What does this value represent? Certainly we have seen that in Lindsey's model, some HRs are worth less than 1.42 runs while others are worth more. But on average, for a large sample of situations, Lindsey found that in 1959–1960, a good estimate of the value of a home run was 1.42 runs.

The next question, naturally, is what about the average run values for other plays? Lindsey performed calculations following the same procedure for singles, doubles, and triples. For now, we will just summarize our own numbers, based on Lindsey's calculations, in Table 7-7.[9] On average, the difference between the most and least valuable hits is less than one run. More intriguing are the run value ratios which indicate that a HR is on average about three times as productive as

	Average Run Value	Ratio of Run Value to Run Value of Single	Base Value
Single	0.454 Runs	1	1 Base
Double	0.818 Runs	1.80	2 Bases
Triple	1.066 Runs	2.35	3 Bases
Home Run	1.419 Runs	3.13	4 Bases

TABLE 7-7   Average Values of Hits with Respect to Runs and Bases

a single. For the sake of comparison, the rightmost column presents the relative values of these hits in terms of bases. We see that a base interpretation of value appears to overestimate the run value of extra-base hits. Besides the home run, a triple is worth less than three times the runs of a single, and a double is worth less than twice the runs of a single.

Lindsey's analysis provides some insight into why the Slugging Percentage (SLG) provides a relatively poor estimate of team run production. As discussed in the previous chapter, SLG is just total bases divided by at-bats, where total bases uses the 1:2:3:4 ratios for singles, doubles, triples, and home runs. Perhaps a modified version of SLG which uses the average run values for hits to determine average runs per at bat (instead of SLG's average bases per at bat) would be an improvement.

Lindsey calculated run productivity using the following formula, where each hit is weighted by its average run value from Table 7-7:

$$Average\ runs\ per\ at\ bat = \frac{(.454 \times 1B) + (.818 \times 2B) + (1.066 \times 3B) + (1.419 \times HR)}{AB}$$

When Lindsey's estimate of average runs per at-bat is used to estimate team runs per game in the years 1876–1999, its RMSE is lower than SLG's RMSE in 108 out of the 124 years. So, simply modifying the 1:2:3:4 weights to .454:.818:1.066:1.419 provides a significant improvement in estimating team runs.

Palmer Enters the Picture

Pete Palmer has been a consultant for the official statisticians of the American League, chairman of SABR's statistical analysis committee, and an editor of baseball encyclopedias including the current standard, *Total Baseball*. From the

[9] Although we used his procedures for calculation, the results in Table 7-7 are slightly different from Lindsey's 1963 run values: single (.41), double (.82), triple (1.06), and home run (1.42).

mid-1960s into the early 1980s, Palmer conducted his own research on evaluating offensive performance building upon the foundation established by Lindsey. Palmer moved this work forward on two fronts:

1. He expanded the model beyond hits to include walks, hit by pitcher, steals, caught stealing, and outs. For each of these plays, he calculated the average number of runs added or (in the case of caught stealing and outs, subtracted) by the play.

2. He developed a computer simulation to model run production through baseball history. The simulation allowed him to replace Lindsey's data in Table 7-4 (taken from a relatively small set of games from only two years) with separate tables of run production in all base-out initial states for different periods of time. Surprisingly, as shown in Figure 7-8, Palmer found little variability in the average run values of different plays across the decades of baseball in the twentieth century. (The one exception is outs, which we will address shortly.)

Based on these results, Palmer settled on the following estimate of runs scored, which he called the *Linear Weights formula*:

$$LWTS = (.46 \times 1B) + (.80 \times 2B) + (1.02 \times 3B) + (1.40 \times HR) +$$
$$[.33 \times (BB + HBP)] + (.30 \times SB) + (-.60 \times CS) + [-.25 \times (AB - H)]$$

A comparison of the Linear Weights model with Lindsey's model shows very little difference in the average runs for the various hits. Because neither a walk nor a hit by pitcher has the advancement capability of a hit, they have less run value than a single.

One major feature of Palmer's model that is not part of Lindsey's is Palmer's inclusion of outs from plate appearances (as estimated by AB – H) and outs from being caught stealing. These plays, which have negative run values, make it possible for LWTS to have a value *below zero* if not enough positive plays (hits and walks) are accumulated to offset the expected runs lost from outs. So, LWTS does not estimate the total number of runs produced. Instead, it estimates the number of runs produced *above the average expected* for the number of play events.

For example, let's consider the offensive records of two New York shortstops in 1999. The Mets' Rey Ordonez was defensively brilliant, but offensively challenged. Derek Jeter of the Yankees, though maybe not quite as good defensively, was a thoroughly superior offensive player. Table 7-8 presents their offensive statistics for 1999 as well as their Linear Weights ratings. According to Palmer's

Player	AB	H	1B	2B	3B	HR	SB	CS	BB	HBP	LWTS
Derek Jeter	627	219	149	37	9	24	19	8	91	12	73.81
Rey Ordonez	520	134	107	24	2	1	8	4	49	1	–8.14

TABLE 7-8   Run Production by Derek Jeter and Rey Ordonez in 1999 as Estimated by LWTS

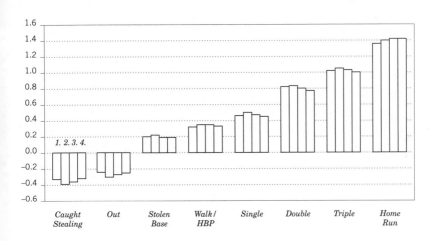

1.   1901–1920
2.   1921–1940
3.   1941–1960
4.   1961–1977

FIGURE 7-8   Average run values for different plays in different periods of baseball history (Palmer Simulation).

model, Jeter contributed many more runs (almost 74 more) than expected by an average player in a season while Ordonez contributed about 8 less than the average player.

As we saw in Figure 7-8, the run value of an out has fluctuated throughout baseball history. Palmer uses the run value of an out to adjust Linear Weights ratings for the average productivity of players. So, the run value of an out is larger in years when run production is high (–.30 from 1921–1940) and smaller in periods of low run production (–.24 from 1901–1920). This makes sense, since each out decreases the opportunity to score more runs, and thus is more damaging in eras when you expect to score a lot than in eras when runs are harder to come by.

Whether or not you believe in the juiced ball, it is indisputable that 1999 was a year with high run production throughout the major leagues. So, the run value of an out was closer to –.30 runs than to Palmer's standard value of –.25. Using –.30 instead of –.25, then, we obtain LWTS ratings of –27.44 and +53.41 runs for Ordonez and Jeter respectively. So, after adjusting for run production during the year, we find that Jeter was actually closer to the average 1999 player, while Ordonez was even farther from average than originally thought.[10] In essence, the Linear Weights formula as given above with a –.25 run value for outs estimates runs above average, where "average" represents an historical standard, not average with respect to a particular year such as 1999.

Palmer's modifications to Lindsey's ideas do improve estimates of team runs scored. LWTS, like Runs Created, provides a cumulative (actually, net) estimate of runs produced. So, as with RC, we must divide it by games before correlating it with team run production data. We found that the RMSE for LWTS/G is less than that of Lindsey's model in 45 of the last 46 years of baseball history. Since the weights for hits are only slightly different, the improvement in fit is due mainly to the inclusion of walks, hit by pitcher, and stolen bases in the model.

Comparing the LSLR and Lindsey-Palmer Models

Practically speaking, the regression model LSLR and the Lindsey-Palmer model are basically the same. Both models assign average run values to each play. The difference lies in the techniques used to find those values or weights. The LSLR model weights were found using the standard statistical techniques of linear regression based on team data from 1954–1999. Lindsey and Palmer found their weights empirically—by analyzing the changes in large numbers of actual baseball game situations, and the results produced by each play type.

Despite the different paths taken by each model, LSLR and the Lindsey-Palmer models arrive at very similar weights. Using the Palmer version of LWTS, Table 7-9 compares the models. The weights for singles and walks are close matches. The LSLR model gives less weight to doubles and greater weight to triples and home runs. The difference in the doubles weight is especially large. There are several possible reasons for this. One is that the base-running assumptions used to generate the changes in base situations for the LWTS

---

[10] Using the formula with a –.3 run value for outs, Chipper Jones of the Atlanta Braves had the highest 1999 LWTS rating of 72.20 runs above average while Mike Caruso of the Chicago White Sox had the lowest LWTS rating –47.73 runs below average. Of Caruso's 132 hits in 529 at bats, only 17 were for extra bases. Ordonez had the sixth lowest LWTS rating in 1999, while Jeter had the seventh highest.

Weight (Runs)	w_{SB}	w_{1B}	w_{2B}	w_{SF}	w_{3B}	w_{HR}	w_{BB}
LWTS	.30	.46	.80	•	1.02	1.40	.33
LSLR with SF	.14	.49	.61	.73	1.14	1.50	.33

TABLE 7-9   Model Weights for LSLR (with Sacrifice Flies) and LWTS

model are too liberal. However, a calculation of the Lindsey weight, assuming just a 2-base advance, only reduces the double weight to .76 runs. Another possibility is that the LSLR weight for home runs has been increased at the expense of the double weight because of the correlation between HRs and doubles. A third possibility is that the weights reflect additional information carried by the frequencies of plays. The LWTS weights reflect only the value of the play itself, since they are constructed by calculating the change in run production produced by each play. The linear regression technique used to develop the LSLR weights only discovers *overall tendencies* in run production as the number of play events changes. This may be good or bad. It's bad in the sense that we are not sure exactly what each measure represents within the LSLR model. It's good in the sense that the LSLR model may capture aspects of baseball within the data that are not measured explicitly.

This is related to the question about sacrifice flies that initiated our investigation of the Lindsey-Palmer approach. What is the true value of a sacrifice fly in terms of runs? Notice that we do not have a LWTS weight for sacrifice flies in Table 7-9. However, we can use Lindsey's data in Table 7-4 to calculate one.

We do this in the same way that we evaluated the value of a home run in Table 7-5. In fact, it is easier. A home run can occur in all 24 base-out situations, but a sacrifice fly can only occur in 8 base-out situations, those with less than two outs and a runner on third base.[11] A sacrifice fly guarantees that a run will score and at least one out will occur. We have assumed that only one out occurs and only the runner on third advances. So, our calculation, shown in Table 7-10, is much less complex than the HR calculation.

Looking at the "Change in Runs" column, we see that a sacrifice fly produces an increase in expected runs only with one out and a runner on third, or runners

[11] A runner on third base is not required for a sacrifice fly. A batter can be awarded a sacrifice fly for scoring a runner from any base. For example, on April 3, 2001, the Phillies' Brian L. Hunter scored the game-winning run from second base on a long fly out by Doug Glanville, who was awarded a sacrifice fly. However, such sacrifice flies with no runner on third base are so rare that our calculation is unaffected by ignoring them.

INITIAL STATE				FINAL STATE				CHANGE
Bases Occupied	Outs	# of Situations	Expected Runs	Bases Occupied	Outs	Expected Runs	Runs	Total Runs	Runs
3	0	67	1.39	None	1	.243	1	1.243	−.147
	1	202	0.980	None	2	.102	1	1.102	.122
1, 3	0	119	1.940	1	1	.498	1	1.498	−.442
	1	305	1.115	1	2	.219	1	1.219	.104
2, 3	0	73	1.960	2	1	.671	1	1.671	−.289
	1	176	1.560	2	2	.297	1	1.297	−.263
Full	0	92	2.220	1, 2	1	.939	1	1.939	−.281
	1	215	1.642	1, 2	2	.403	1	1.403	−.239

TABLE 7-10   The Run Values of a Sacrifice Fly

on first and third. In the worst case, a sacrifice fly with no outs and runners on first and third loses almost half a run on average. Weighting each change in runs by the relative frequency of the situation, we find that the average value of a sacrifice fly in runs is about −.12. This is far different from the result of .73 runs found by the regression techniques for the LSLR model. We conclude that the LSLR weight for sacrifice flies probably captures the high expected runs value of the state in which it occurs. Note that the expected runs for the initial state in Table 7-10 range from a minimum of .98 runs to a maximum of 2.22. This is high relative to the expected runs for all 24 initial states.

It appears, then, that we have found a case where linear regression may not provide a useful weight for evaluating player performance. Even though the inclusion of SF with a .73 run weight reduces the RMSE, it is not advisable to include the weight in the LSLR model. The lesson here is that while reducing error (RMSE) is the major objective, it should not be done at the expense of creating a model that does not have some common sense built into it as well. The results of regression analysis should not be accepted without questioning model assumptions, the data used for the analysis, and ultimately the reasonableness of the answers.

One characteristic that both techniques have in common is their reliance on data for the development of the model. Regression techniques require a set of data describing the value to be estimated and the quantities used in the estimation. In our case with LSLR, the data set contained runs scored and the number of various play events per team per season from 1954–1999. For Lindsey, the data set was detailed play-by-play results from games in the 1959–1960 seasons. For Palmer, the data sets were the frequencies of various plays in each year, which were used to drive his computer simulation.

Is it possible to break our dependence on data to develop a run production model? In some ways, the intuitive models examined in Chapter 6 did this. None of those models was developed using data. Each was inspired by a theory about what contributed to run production. However, here we have something different in mind. We would like to use principles of probability to build a model based on how the game is played.

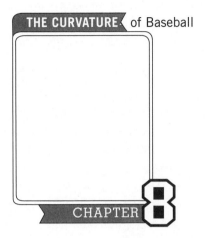

THE CURVATURE of Baseball

CHAPTER 8

One feature that makes baseball lend itself to statistical analysis is the *discrete* nature of the game. Every inning unfolds play by well-defined play, and at the conclusion of each play, the inning must be in 1 of 25 well-defined states. We can break down these states into 8 distinct base situations:

None on.

Runner on first.

Runners on first and second.

Runners on first and third.

Runners on first, second, and third (bases loaded).

Runner on second.

Runners on second and third.

Runner on third.

We can also break down the states into 3 distinct out situations (0, 1, and 2 outs). We combine these 3 out and 8 on-base possibilities to get 24 states ($3 \times 8 = 24$), then add 1 for the final state of the inning (3 outs), giving us a total of 25. Every situation before and after a play in baseball must fit into one, and only one, of these 25 unambiguous and clearly distinguished states.

Compare this with other sports. In football, we could consider one team's possession as equivalent to a team's inning at bat. How many possible states are there at the end of each play in a single possession? With some simplification, there are 100 different positions on the field, one for each yard mark. Then there are 4 different downs. And then there are 10 different yards to go (for first down). If we multiply these values, we obtain a large number

Field positions × downs × yards to go = possible states

$$100 \times 4 \times 10 = 4000$$

The number 4000 may seem large, but it is a lower bound—that is, it's extremely understated. The number of yards to go for a first down may be greater than 10, and both this value and the field position value (the line of scrimmage) are actually continuous. In other words, the ball is rarely placed exactly at a yardline mark, so neither of these values has to be an integer. Any fractional value for yards to go, or any line of scrimmage between yard markers, would also have to be considered, thus making the number of possible states virtually infinite.

So there are at least 160 (or 4000/25) possible states in football for every comparable base-out state in baseball. And football is relatively discrete compared to other major sports, like basketball, hockey, and soccer, which have a continuous flow in both time and space. The number of possible situations these other sports present is literally infinite, and in the course of each game or match, there are very few moments where the action is paused in an easily defined and numerically described state.

The D LS I Simulation Model

Several researchers besides Lindsey and Palmer have taken advantage of baseball's relatively simple, static, and discrete structure to create probabilistic models of run production. Among the earliest was the Scoring Index model developed by D. A. D'Esopo and B. Lefkowitz in a 1960 SRI internal report. Their work was not available publicly until it was published in the groundbreaking collection of papers *Optimal Strategies in Sports* in 1977.[1] In the interim, essentially the same model was developed independently by Thomas M. Cover and Carroll W. Keilers, who used it to create a batting statistic called the Offensive Earned Run Average (OERA).

At its heart, the *D'Esopo-Lefkowitz Scoring Index (DLSI)* model starts with the basic premise of the BRA, DX, and RC/G models: scoring runs is the product of two related types of events, getting on base and advancing runners. However, their model puts a different spin on this premise, separating run production into two processes:

---

[1] This is an excellent work, now out of print but probably still obtainable in many college libraries. The techniques presented by many papers in the collection are as relevant today as when they were published more than two decades ago.

1. *Getting on base*: An event that describes getting on base a
   particular number of times in an inning.

2. *Advancing around the bases*: An event that describes scoring a
   given number of runs when a particular number of players get on
   base in an inning.

For ease of computation and explanation, we'll make some basic assumptions
here about the types of events that occur in a plate appearance and how runners
advance on the bases. The variant of the DLSI model we use assumes that the
only possible events in a plate appearance are BB/HBP, 1B, 2B, 3B, HR, and out.
All outs are effectively strikeouts (i.e., single outs leaving runners ·in place).
A single scores runners on second and third bases, but only advances a runner
on first to second base. A double advances all runners two bases. Although you
might quibble with this choice of rules, we will see that this model gives results
pretty similar to real baseball.

The Probability of Scoring Two Runs

Using the DLSI model, suppose that we are interested in the following event:

{exactly 2 runs score in an inning}

We need to think of all of the possible ways for runners to reach base so that 2
runs can be scored. We will explain shortly that 2 runs can score when 2, 3, 4, or 5
runners reach base in the inning. So if for now we take that as a given, we can
break down the event {exactly 2 runs score in an inning} into the following events:

{2 players reach base in inning, and exactly 2 runs score}

or

{3 players reach base in inning, and exactly 2 runs score}

or

{4 players reach base in inning, and exactly 2 runs score}

or

{5 players reach base in inning, and exactly 2 runs score}

The tree diagram in Figure 8-1 illustrates the two steps of this process: (1) players reach base, and (2) players score 2 runs.

To compute the probability of scoring 2 runs in an inning, we first assign probabilities to the branches of the tree diagram. At the first set of branches, we compute the probabilities that 2 players reach base, 3 reach base, 4 reach base, and 5
reach base. We will call these probabilities Pr(2 reach base), Pr(3 reach base), and

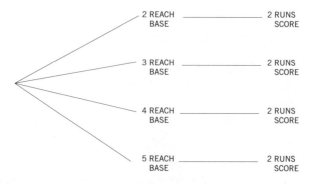

FIGURE 8-1   Tree diagram for scoring exactly 2 runs in an inning.

so forth. Then, at the second set of branches, we find the probability of scoring 2 runs *if* 2 players reach base, which we will denote Pr(2 runs | 2 on base), the probability of scoring 2 runs if 3 reach base Pr(2 runs | 3 on base), and so forth. We place these probabilities on the branches of the tree diagram in Figure 8-2.

After all of the probabilities of the branches have been assigned, we find the probability of scoring 2 runs by multiplying probabilities along each branch of the tree, then summing the products:

$$\text{Pr(2 runs)} = \text{Pr(2 reach base)} \times \text{Pr(2 runs | 2 reach base)} +$$

$$\text{Pr(3 reach base)} \times \text{Pr(2 runs | 3 reach base)} +$$

$$\text{Pr(4 reach base)} \times \text{Pr(2 runs | 4 reach base)} +$$

$$\text{Pr(5 reach base)} \times \text{Pr(2 runs | 5 reach base)}$$

Let's back up and explain some of our logic. In order to score 2 runs, at least 2 players have to reach base. After all, if no players get to first base, how can *any* runs score? And if only 1 player reaches first base, the player may or may not score, but at most 1 run will score. In general, if we want to find the probability of scoring R runs (where $R = 2$ in this setting), we only have to look at innings in which *at least* R players get on base.

Next, we see that to find the probability of 2 runs scoring we don't have to look at innings where 6 or more batters get on base. Why is this? Let's look at the case where 6 players reach first base on walks. Assuming that no players are caught stealing, hit into double plays, or are thrown out at a base, 3 runs will score in this inning; the first 3 walks will load the bases and each of the next 3 will force in a run. Because of our assumptions about outs, the order in which outs are interspersed with the walks does not matter—3 runs will still score in this inning. And this is the most conservative estimate! The walk (or hit by

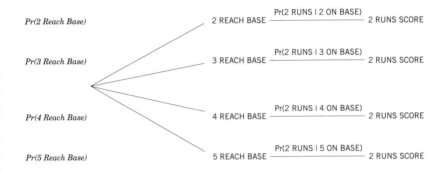

FIGURE 8-2   Tree diagram for scoring exactly 2 runs in an inning
(expanded with probabilities of scoring for different numbers of players
reaching base).

pitcher) is the least productive of on-base events. If we substitute any type of hit
(single, double, triple, or home run) for one of the walks, more than 3 runs are
liable to score. So, we can say that if 6 players reach base in an inning, at least 3
runs will score. Consequently, if we wish to compute the probability of scoring
exactly 2 runs, we don't have to consider any innings in which 6 or more players
get on base.[2] In general, if we want to find the probability of scoring R runs, we
can ignore innings in which *at least* $R + 4$ players get on base.

This is why in finding the probability of scoring a specific number of runs R
in an inning, we have only to look at four cases: innings in which $R, R + 1, R + 2$,
and $R + 3$ players get on base.

The Probability of Scoring No Runs

So what is the probability of scoring R runs in each of these cases? Let's examine the simplest case, where $R = 0$ (*no* runs score in the inning).

We first focus on computing probabilities at the second set of branches of the
tree; that is, the probability of scoring 0 runs given different number of players
on base. The first value we need is the probability of scoring 0 runs when 0 players get on base. An easy question: the probability is 1 because if no players reach
base, no runs can score:

$$Pr(0 \text{ runs} \mid 0 \text{ reach base}) = 1$$

[2] Note that the model does not consider the case where a batter reaches first base on a fielder's
choice as an on-base event.

How about the probability of scoring 0 runs when 1 player gets on base? Assuming no stolen bases (a basic assumption of the model), there is only one way that a run can be scored with one on-base event; that is, if the event is a home run. So, the probability of no runs scored (when 1 player gets on base) equals 1 minus the fraction of on-base events that are home runs. The fraction of on-base events that are home runs, denoted by f_4, is defined as follows:

$$f_4 = \frac{HR}{BB + HBP + 1B + 2B + 3B + HR}$$

So,

$$\Pr(0 \text{ runs} \mid 1 \text{ reaches base}) = 1 - f_4$$

While we are at it, let's define proportions for all on-base events that are walks / hits by pitcher, singles, doubles, and triples:[3]

$$\textit{Fraction of walks / hit by pitcher } f_0 = \frac{BB + HBP}{BB + HBP + 1B + 2B + 3B + HR}$$

$$\textit{Fraction of singles } f_1 = \frac{1B}{BB + HBP + 1B + 2B + 3B + HR}$$

$$\textit{Fraction of doubles } f_2 = \frac{2B}{BB + HBP + 1B + 2B + 3B + HR}$$

$$\textit{Fraction of triples } f_3 = \frac{3B}{BB + HBP + 1B + 2B + 3B + HR}$$

Only two more cases to go, but they are the most difficult ones. Let's consider the probability of scoring 0 runs when 2 players get on base:

- When the first player to get on base gets a walk or hit by pitcher, the following sequences do not score a run: *BB–BB, BB–1B, BB–2B*.[4] The probability of these sequences is $f_0 (f_0 + f_1 + f_2)$.

- Similarly, when the first player to get on base gets a single, the following sequences do not score a run: *1B–BB, 1B–1B, 1B–2B*. The probability of these sequences is $f_1 (f_0 + f_1 + f_2)$.

---

[3] We use the notation f_n where n indicates the number of bases for the hit involved and $n = 0$ for a walk or hit by pitcher

[4] To simplify, we will use BB to symbolize a walk or hit by pitcher. Both have the same effect.

- When the first player to get on base gets a double or triple, only two sequences do not score a run: *2B–BB, 3B–BB*. The probability of these sequences is $(f_2 + f_3) f_0$.

To summarize, if 2 players get on base, the probability of not scoring any runs is:

$$\text{Pr}(0 \text{ runs} \mid 2 \text{ reach base}) = f_0 (f_0 + f_1 + f_2) + f_1 (f_0 + f_1 + f_2) + (f_2 + f_3) f_0$$

$$= (f_0 + f_1)(f_1 + f_2) + (1 - f_4) f_0$$

We are down to the last case, the probability of scoring no runs when 3 players get on base. We can calculate this value by enumerating all the sequences as we did above. It turns out when you do this (consider this an at-home exercise), the non-scoring sequences are the ones above for 2 players on base with a walk or HBP appended at the end (e.g., *BB–2B* becomes *BB–2B–BB*). So, the probability of scoring 0 runs if three players get on base is just Pr(0 runs | 2 reach base) times the fraction of on-base events that are walks / HBP:

$$\text{Pr}(0 \text{ runs} \mid 3 \text{ reach base}) = \text{Pr}(0 \text{ runs} \mid 2 \text{ reach base}) \times f_0$$

The probability of scoring no runs (similar to the tree diagram for scoring 2 runs in Figure 8-2) is the weighted sum of these probabilities: Pr(0 runs | 0 reach base), Pr(0 runs | 1 reaches base), Pr(0 runs | 2 reach base), and Pr(0 runs | 3 reach base). The weights are just the probabilities of putting the respective players on base: Pr(0 reach base), Pr(1 reaches base), Pr(2 reach base), and Pr(3 reach base). So,

$$\text{Pr}(0 \text{ runs}) = \text{Pr}(0 \text{ reach base}) \times \text{Pr}(0 \text{ runs} \mid 0 \text{ reach base}) +$$

$$\text{Pr}(1 \text{ reaches base}) \times \text{Pr}(0 \text{ runs} \mid 1 \text{ reaches base}) +$$

$$\text{Pr}(2 \text{ reach base}) \times \text{Pr}(0 \text{ runs} \mid 2 \text{ reach base}) +$$

$$\text{Pr}(3 \text{ reach base}) \times \text{Pr}(0 \text{ runs} \mid 3 \text{ reach base})$$

In order to complete this calculation, all we need is a formula for the probability that a given number of players get on base. (These will be the probabilities at the first set of branches in our tree diagram.) Fortunately, this is a well-understood statistical process and can be calculated quite simply as follows:[5]

---

[5] Many will recognize this as a negative binomial distribution for the number of players who get on base before 3 outs occur. The simplified model assumes that all outs occur from batters, not runners. A batter is either out or safe on base.

$$\text{Pr(B reach base)} = \frac{(B + 2)\,(B + 1)\,p^B\,(1 - p)^3}{2}$$

where B is the number of players who get on base and p is the probability that a batter gets on base in a plate appearance. So, the probabilities we need are

$$\text{Pr(0 reach base)} = \frac{(0 + 2)\,(0 + 1)\,p^0\,(1 - p)^3}{2} = (1 - p)^3$$

$$\text{Pr(1 reaches base)} = \frac{(1 + 2)\,(1 + 1)\,p^1\,(1 - p)^3}{2} = 3p\,(1 - p)^3$$

$$\text{Pr(2 reach base)} = \frac{(2 + 2)\,(2 + 1)\,p^2\,(1 - p)^3}{2} = 6p^2(1 - p)^3$$

$$\text{Pr(3 reach base)} = \frac{(3 + 2)\,(3 + 1)\,p^3\,(1 - p)^3}{2} = 10p^3(1 - p)^3$$

Clearly, a reasonable estimate for p to use in calculations is our old friend the team on-base percentage.[6] Based on our probability formula for B, we can estimate the average number of players to reach base in an inning $(\bar{B})$ as follows:[7]

$$\textit{Average reaching base in an inning: } \bar{B} = \frac{3p}{1 - p}$$

Through some clever insight into the sequence of hits and walks that produce runs, D'Esopo and Lefkowitz found that the average number of runners left on base in their model is

$$\bar{L} = \text{Pr(0 runs} \mid \text{1 reaches base)} \times [1 - \text{Pr(0 reach base)}] + $$

$$\text{Pr(0 runs} \mid \text{2 reach base)} \times [1 - \text{Pr(0 reach base)} - \text{Pr(1 reaches base)}] + $$

$$\text{Pr(0 runs} \mid \text{3 reach base)} \times [1 - \text{Pr(0 reach base)} - $$

$$\text{Pr(1 reaches base)} - \text{Pr(2 reach base)}]$$

---

[6] D'Esopo and Lefkowitz used a slightly more involved estimate for p. Their estimate was also based on OBP, but it added errors and subtracted double plays from the numerator. In addition, it added sacrifice flies to the denominator.

[7] A bar over a symbol is often used to denote the average value, as in the average number of base runners in an inning here.

Innings	AB	BB	HBP	1B	2B	3B	HR
11,047	42,330	3974	232	7744	1788	324	1159

TABLE 8-1   1959 National League Data

The amazing thing about this model is that these formulas used to calculate the probability of *not scoring* provide you with all the tools you need to estimate the average number of runs scored per inning. The basic principle behind this result is the Law of Batter Conservation:

> Every batter is either out, scores, or is left on base.

Since the model assumes that only batters are out and that runners either score or are left on base, the average number of runs scored in an inning $(\bar{R})$ is just the average number of players who get on base $(\bar{B})$ minus the average number of runners left on base $(\bar{L})$ in an inning. D'Esopo and Lefkowitz called their estimate of the average number of runs scored per inning the Scoring Index, which (as mentioned earlier) we abbreviate as DLSI.

A DLSI Example

Maybe we can get a better handle on this simulation model if we perform a sample calculation. The year 1959 was one of the most interesting in baseball history. It was the only year in the decade from 1955 through 1964 that the Yankees were *not* in the World Series. The Dodgers, in only their second year in Los Angeles, took advantage of the Yankees' absence to win the series with a team generally regarded as one of the weakest of all World Champions. The greatest part of the challenge for the Dodgers was to defeat the powerful Milwaukee Braves in a single National League playoff game after the conclusion of the regular 154-game season.[8] For our purposes, 1959 marks the beginning of serious baseball run production models by Lindsey, D'Esopo, and Lefkowitz, so in their honor we'll use the 1959 National League for our example. We start with Table 8-1, which provides totals for the National League in 1959.

Run production can be characterized by the probability p of getting on base estimated by the following:

---

[8] See the essay on manager Fred Haney and the 1959 Milwaukee Braves in *The Bill James Guide to Baseball Managers* for a wonderful description of how such a great team as the Braves of this period could have been upset in their quest for a third consecutive National League title.

$$p = \frac{BB + HBP + 1B + 2B + 3B + HR}{AB + BB + HBP}$$

$$= \frac{3974 + 232 + 7744 + 1788 + 324 + 1159}{42330 + 3974 + 232} = .32708$$

and by the proportions of on-base events (walks, singles, etc.):

$$f_0 = \frac{BB + HBP}{BB + HBP + 1B + 2B + 3B + HR} = \frac{4206}{15{,}221} = .27633$$

$$f_1 = \frac{1B}{BB + HBP + 1B + 2B + 3B + HR} = \frac{7744}{15{,}221} = .50877$$

$$f_2 = \frac{2B}{BB + HBP + 1B + 2B + 3B + HR} = \frac{1788}{15{,}221} = .11747$$

$$f_3 = \frac{3B}{BB + HBP + 1B + 2B + 3B + HR} = \frac{324}{15{,}221} = .02129$$

$$f_4 = \frac{HR}{BB + HBP + 1B + 2B + 3B + HR} = \frac{1159}{15{,}221} = .7614$$

These are the basic elements used by the model. Batters got on base about 32.7 percent of the time, and of the times that they got on base, 27.6 percent were via a walk or hit by pitch, 50.9 percent via a single, 11.7 percent via a double, 2.1 percent via a triple, and 7.6 percent via a home run.

The most complicated part is the three-step procedure used to calculate the average number of runners left on base per inning:

1. We calculate the probability of *not* scoring when 1, 2, or 3 runners reach base:

$$\Pr(0 \text{ runs} \mid 1 \text{ reaches base}) = 1 - f_4 = 1 - .076 = .924$$

$$\Pr(0 \text{ runs} \mid 2 \text{ reach base}) = (f_0 + f_1)(f_1 + f_2) + (1 - f_4)f_0$$

$$= (.276 + .509)(.509 + .117) + .924\,(.276)$$

$$= .747$$

$$\Pr(0 \text{ runs} \mid 3 \text{ reach base}) = \Pr(0 \text{ runs} \mid 2 \text{ reach base}) f_0$$
$$= .747 \times .276$$
$$= .206$$

As we might expect, the probability of not scoring at all drops from 92 percent to 75 percent to 21 percent as more runners reach base.

2. We calculate the probabilities that 0 batters reach base, exactly 1 batter reaches base, and exactly 2 batters reach base:

$$\Pr(0 \text{ reach base}) = (1-p)^3 = (1-.327)^3 = .305$$

$$\Pr(1 \text{ reaches base}) = 3p\,(1-p)^3 = 3 \times .327\,(1-.327)^3 = .299$$

$$\Pr(2 \text{ reach base}) = 6p^2\,(1-p)^3 = 6 \times .327^2\,(1-.327)^3 = .196$$

3. We then use these values to calculate the average number of runners left on base per inning:

$$\bar{L} = \Pr(0 \text{ runs} \mid 1 \text{ reaches base}) \times [1 - \Pr(0 \text{ reach base})] + \Pr(0 \text{ runs} \mid 2 \text{ reach base}) \times$$
$$[1 - \Pr(0 \text{ reach base}) - \Pr(1 \text{ reaches base})] + \Pr(0 \text{ runs} \mid 3 \text{ reach base}) \times$$
$$[1 - \Pr(0 \text{ reach base}) - \Pr(1 \text{ reaches base}) - \Pr(2 \text{ reach base})]$$

3. or
$$\bar{L} = (.924 \times [1 - .305]) + (.747 \times [1 - .305 - .299]) +$$
$$(.206 \times [1 - .305 - .299 - .196]) = .979$$

The last step is easy. Using p, we can immediately calculate the average number of runners reaching base per inning:

$$\bar{B} = \frac{3p}{1-p} = \frac{3 \times .32708}{1 - .32708} = 1.458$$

The average number of runs scored per inning is just the average reaching base minus the average left on base. Therefore:

$$DLSI = \bar{R} = \bar{B} - \bar{L} = 1.458 - .979 = .479 \text{ runs per inning}$$

is the estimated average number of runs scored per inning by National League teams in 1959.

Notice that in the course of this calculation, the simulation model required the computation of other values, such as the average number of runners left on base, various probabilities of putting runners on base, and probabilities of scoring with runners on base. This makes the calculation somewhat longer than other models, but it does provide the benefit of an added richness to our understanding of the game. Using these results in some additional calculations, we can compute the distribution of runs scored per inning, that is, the probability of scoring. For example, recall that the probability of not scoring in an inning is calculated as follows:

$$\text{Pr(0 runs)} = \text{Pr(0 reach base)} \times \text{Pr(0 runs | 0 reach base)} +$$

$$\text{Pr(1 reaches base)} \times \text{Pr(0 runs | 1 reaches base)} +$$

$$\text{Pr(2 reach base)} \times \text{Pr(0 runs | 2 reach base)} +$$

$$\text{Pr(3 reach base)} \times \text{Pr(0 runs | 3 reach base)}$$

We have already calculated most of these values. The only extra value we need to calculate is:

$$\text{Pr(3 reach base)} = 10p^3 (1 - p)^3 = 10 \times .327^3 (1 - .327)^3 = .107$$

Substituting these values, we obtain:

$$\text{Pr(0 runs)} = (.305 \times 1) + (.299 \times .924) + (.196 \times .747) + (.107 \times .206) = .750$$

So, we might expect that no runs were scored in about 75 percent of the innings played by National League teams in 1959.

D'Esopo and Lefkowitz used such a calculation as a test of their model. They calculated a distribution of runs scored per inning with their model and compared the result against Lindsey's data and against similar data they collected from 100 games in the 1959 National League baseball season. Table 8-2 shows the data they collected as well as predictions based on our version of their model. The agreement is quite good, considering the relatively simple assumptions of the simulation model (no stealing, no bunting, no advancement on outs).

Table 8-2 also shows the average number of runs scored per inning in the data together with the model prediction. In fact, a total of 5462 runs were scored in the 1959 National League season over 11,047 innings, for an average of .494 runs per inning. The simulation model prediction is somewhat lower than the averages from the data.

DATA

Runs Scored	Lindsey	D'Esopo and Lefkowitz	DLSI Model
0	73.0%	74.4%	75.0%
1	14.6%	12.9%	12.6%
2	7.0%	6.8%	6.7%
3	2.9%	2.9%	3.2%
4	1.4%	2.1%	1.5%
5	0.7%	0.7%	0.6%
6 or more	0.4%	0.3%	0.5%
Average Runs	0.488	0.489	0.479

TABLE 8-2   Runs Scored Per Inning in 1959 National League Season (DLSI Model Results Compared with Data)

Lessons from the Simulation

You might at this point ask, if you haven't asked already, what makes this a *simulation* model? After all, the calculation is similar to that for the other models we have covered (for example, Total Average or Runs Created), except DLSI is more complicated. And you might comment that you thought a baseball simulation was a computer program that played many games over and over to produce results that replicated actual game play. While computer programs are the most common form of simulation today, you should recall that APBA, Strat-O-Matic, and Sports Illustrated—the board games we discussed in Chapter 1—are simulations too. These games do not require a computer, although if you want to simulate many entire seasons, you would have to use a computer version.

The D'Esopo-Lefkowitz model differs from these board games only in the relative simplicity of its rules and its assumption of a single average level of performance for all hitters. (What distinguishes the D'Esopo-Lefkowitz model from the other models reviewed is its genesis from the rules of baseball applied in a probabilistic way.) But we could use the rules they define as well as any assumptions we wish for the probability of getting on base and on-base profiles to actually play games of baseball either as a board game or as a computer program.

We could, for example, construct a game very similar to All-Star Baseball. The game would be a much simpler one, consisting of a single disk. The disk would have six slices: walk, single, double, triple, home run, and out. All batters would use this disk. If the game were to simulate play during the 1959 National

League baseball season, the sizes of the slices would be determined by the data we just discussed. Recall that the probability of getting on base was $p = .32708$. So, according to the rules established by the assumptions of D'Esopo and Lefkowitz, the probability of getting an out is $1 - p = .67292$. The out slice would span an arc of $.67292 \times 360 = 242$ degrees. The remaining 118 degrees would be divided into five slices for walks and the various hits in accordance with the values $f_0, f_1, f_2, f_3,$ and f_4. For example, the home run slice would be:

$$f_4 \times 118 = .07614 \times 118 = 9 \text{ degrees}$$

If we played this game using the rules for runner advancement assumed by D'Esopo and Lefkowitz for the equivalent of many seasons of virtual play (completing all innings until three outs are recorded), we would obtain results for the runs scored per inning which would *exactly* match those estimated using the equations described. The simplicity of the model's rules and player ability assumptions allows us to circumvent the whole process of replaying every plate appearance in every game. The results of playing out the simulation can be obtained simply through calculation, using a few formulas. This is the strength of the D'Esopo-Lefkowitz model.

The weakness of the model also lies in its assumptions. The same simplicity which allows us to capture the information from thousands of seasons of replays with only a few calculations also means that some richness of detail—from the running game, "small ball" advancement from outs, and variation from different player abilities—has not been included. The D'Esopo-Lefkowitz model simulates baseball with very broad strokes. Nonetheless, as indicated by its distribution of runs per inning in Table 8-2, the simulation produces quite reasonable results despite the simplicity of its rules. The model tends to underestimate Team Runs Scored per Inning, perhaps because of its lack of a more sophisticated set of rules describing the advancement of runners on outs.[9] These could be incorporated into the more general version of the model developed by Cover and Keilers.

Now that we have some familiarity with the mechanics of calculating runs per inning with the model, let's examine some aspects more closely. The model says that there are two key elements in run production:

- The variable p, which states the probability of getting on base
  (and avoiding being out).

---

[9] Using RMSE tests such as those in Chapter 6, we find DLSI to be one of the best models tested. It has smaller RMSEs than the basic Runs Created formula (which uses a comparable set of data), but larger RMSEs than Runs Created Tech-1, which has the advantage of using more data (e.g., stolen bases, caught stealing, sacrifice flies).

- The parameters f_0, f_1, f_2, f_3, and f_4, which we will refer to collectively as the *on-base profile*. This profile describes the distribution of all on-base events considered in the model (walk, hit by pitcher, single, double, triple, and home run), and it always sums to 1.

How much can these elements vary from team to team and from year to year? Considering the modern era of baseball, from 1901 through 1999, a typical value for p is .33, close to the 1959 National League average. The 1908 Brooklyn Dodgers had the lowest p, with .266 batters reaching base per opportunity. Lest you think that such a low probability of getting on base is a phenomenon only of the deadball era, consider that this value was challenged by the New York Mets in 1965 (with $p = .278$). The 1950 Boston Red Sox had the highest ($p = .385$), almost 50 percent higher than the minimum value. Not surprisingly, the quintessential on-base batter, Ted Williams, was on this team, although he played little more than half of the season. And a team as recent as the 1994 New York Yankees had an exceptionally high value ($p = .377$), albeit in a strike-shortened season. So there is, from team to team and season to season, considerable variation in this measure of performance.

Table 8-3 shows the range of values for each component of the on-base profile statistic. The table shows a great deal of diversity within each component. For example, the majority of on-base events for some teams were singles (as high as 69 percent for the 1902 St. Louis Cardinals), while others had less than half (as low as 39 percent for the 1999 Oakland A's). However, since the components for each team must add up to 1, they are not independent of each other; you can't increase one component without decreasing at least one of the others. What we are interested in finding is a realistic *combination* of components which produces the most extreme (low and high) results in run production.

	BB+HBP f_0	1B f_1	2B f_2	3B f_3	HR f_4
Maximum	.378	.685	.177	.069	.118
	1949 AL Phi.	1902 NL St.L.	1997 NL Mon.	1903 AL Bos.	1961 AL NY
Average	.277	.527	.119	.026	.051
Minimum	.174	.391	.072	.005	.002
	1921 NL Phi.	1999 AL Oak.	1902 NL Phi.	1998 AL Balt.	1908 AL Chi.

TABLE 8-3   Range of Team On-Base Profile Values (1901–1999)

To understand how the on-base profile affects run production, we assume that the typical on-base probability p is .33. We calculated the run production of each team from 1901–1999 using its own unique on-base profile, but using the same on-base probability (p = .33) for all teams. We then found the teams with the highest (1947 New York Giants)[10] and the lowest (1908 Chicago White Sox)[11] run production. Since the on-base probability was held constant, the only difference in run production was their on-base profiles. The bar charts in Figure 8-3 compare the on-base profiles of these two teams. Remember, these are not the teams with the highest and lowest run production overall, but the teams that had the best and worst on-base profiles *if the probability of getting on base is kept at .33 for all teams.*

Only 10 percent of the 1908 White Sox on-base events were extra-base hits (not walks or singles), while almost 25 percent of the 1947 Giants on-base events were extra-base hits. As we would expect, the 1947 Giants have a more productive on-base profile than the 1908 White Sox because of the shift of walks and singles to extra-base hits. In fact, when we compare the values in these on-base profiles to the ranges of the individual components in Table 8-3, we see that the

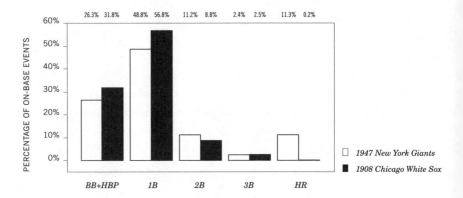

FIGURE 8-3   The best and worst team on-base profiles (1901–1999).

[10] The top on-base profiles included such recent teams as the 1994 Cleveland Indians and the 1997 Seattle Mariners, as well as the legendary 1961 New York Yankees, with Maris and Mantle (which had the highest fraction of home runs, f_4 = 11.8 percent). Apparently, one factor that separates the 1947 Giants from these other high on-base profile teams is their higher fraction of triples, possibly a result of the unusually deep center field of the Giants' home park, the Polo Grounds. The 1990s included some teams with low production on-base profiles; the two lowest were the Los Angeles Dodgers and the Boston Red Sox, both from 1992.

[11] Not only did these White Sox play during the deadball era, but their home field, South Side Park, was "the poorest hitting field in major league history" according to Michael Schell in *Baseball's All-Time Best Hitters.*

profiles are not at all extreme, except in home runs, where the 1947 Giants have one of the highest fractions, while the 1908 White Sox have the lowest.

How does the predicted run production change for our simulation model as either the on-base profile or the probability of getting on base changes? Let's see how the predicted run production will change if we take each of our two extreme on-base profiles, keep them fixed, and change p, the probability of getting on base. Figure 8-4 plots the predicted runs per inning in this scenario. The upper line in this chart shows the predicted runs per inning for the 1947 Giants on-base profile, while the lower line shows the predicted runs per inning for the 1908 White Sox on-base profile. The bullets identify the predicted runs per inning for each team for the actual historical on-base probability p of each team.

The 1947 Giants had an on-base probability near the historical team average, but the 1908 White Sox had an on-base probability way below average. Following the 1908 White Sox line upwards, we see that a team with a poor on-base profile could be as productive as the 1947 Giants if they could compensate for their lack of power with an increase in on-base probability near the historical maximum of .385. Since these are the best and worst on-base profiles, and since the plot encompasses the highest and lowest team p-values from 1901–1999, the predicted run production for *all* teams in the twentieth century lie in the area bounded by these two lines.

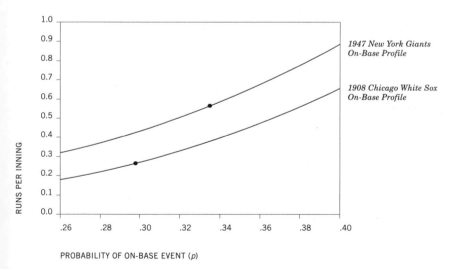

FIGURE 8-4   DLSI for best and worst on-base profiles as the probability of getting on base changes. (Bullets [●] indicate DLSI for team's actual value of p.)

Looking at the 1947 Giants profile, we see that run production does not increase linearly with the probability of getting on base; that is, the line *curves* upward so that run production increases faster as p increases. The 1908 White Sox profile shows a similar (though slightly less pronounced) effect. The vertical distance between the two lines shows the effect of different on-base profiles; that is the effect of varying the distribution of different types of hits. Increasing p increases run production, and improving the on-base profile increases run production, but they improve it in different ways.

DLSI and Runs per Play

Predicted run production for the 1959 National League falls somewhere in the middle of the extremes shown by the Giants and White Sox. Taking another look at the 1959 National League data, what would happen if we added one more walk? That is, keeping everything else the same, how would run production change if the total number of walks were 3975 instead of 3974? Redoing (with greater precision) the calculation described earlier, with this very slight variation, we find that run production would be .4784369 runs per inning, or .0000336 runs per inning higher than with the historical data (.4784033). In 1959, the National League played 11,047 innings. Multiplying this change by the total number of innings gives us the total change in runs

$$.0000336 \times 11047 = .37$$

Doesn't this seem familiar? It looks very close to the run value for a walk or hit by pitcher in the Lindsey, Palmer, and regression models described in Chapter 7. Perhaps the result will continue to match if we do the same thing for singles, doubles, triples, home runs, and outs. Table 8-4 shows the variant data, the predicted change in runs per inning, and the total increase in runs from the play (by multiplying the increase and the number of innings).

Figure 8-5 compares the changes in run production with run values for the same events in the Palmer model. The values are very similar. The biggest difference is in the run value for home runs, which is given more value by the simulation model than by the 1959 NL data. Perhaps this is a reflection of the simulation's assumptions about runner advancement. If the simulation does not allow runners to advance on outs, it may give added weight to home runs, which needless to say are very good at advancing runners. Another possibility is that the increased value of home runs stems from the assumption that runners are never thrown out on the bases. If a walk is followed by a home run, the runner

"WHAT IF" 1959 NATIONAL LEAGUE DATA

RUN CHANGE

Play	AB	BB/HBP	1B	2B	3B	HR	Per inning	Per play
Walk/HBP	42,330	**4207**	7744	1788	324	1159	.0000336	.37
Single	**42,331**	4206	**7745**	1788	324	1159	.0000476	.53
Double	**42,331**	4206	7744	**1789**	324	1159	.0000662	.73
Triple	**42,331**	4206	7744	1788	**325**	1159	.0000895	.99
Home Run	**42,331**	4206	7744	1788	324	**1160**	.0001434	1.58
Out	**42,331**	4206	7744	1788	324	1159	-.0000263	-.29

TABLE 8-4   Change in Run Production in 1959 National League Season When Increasing Each Play Count by 1 (Indicated in Boldface)

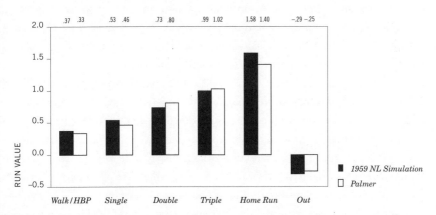

FIGURE 8-5   Run values for plays in the 1959 National League season: DLSI values versus Palmer's Linear Weights values.

will always score; in Palmer's model, he might be erased on a double play or caught stealing. Similarly, the value of a double is certainly reduced by the assumption that doubles never drive in runners from first.[12] Still, the agreement in run values is very good, considering the simplicity of the simulation's assumptions.

The changes in run production predicted by the simulation as we add an additional event to the data is one way of measuring run value per event. Another way of doing this is to calculate the slope of the on-base profiles in graphs like

---

[12] These last two observations—on home runs and doubles—were suggested by David Grabiner in personal correspondence.

Team	Innings	AB	BB	HBP	1B	2B	3B	HR
Cleveland	1458	5634	743	55	1079	309	32	209
Minnesota	1449	5495	500	49	1030	285	30	105

TABLE 8-5   1999 Data for the Cleveland Indians and the Minnesota Twins

the one in Figure 8-4. However, as we observed in Figure 8-4, these on-base profile lines are curved upwards; that is, their slopes increase as the probability of
getting on base increases. Does this mean that plays may have different run values depending on a team's ability to get runners on base?

Let's test this with an example. In 1999, the Cleveland Indians scored more
runs than any other team (1009)[13] while the Minnesota Twins scored the fewest
(686). According to the simulation model, would an extra hit have had more
value to the Indians than it would to the Twins? Table 8-5 gives the 1999 season
data for these two teams.[14] The major differences between them appears to
reside in their abilities to draw walks and hit home runs.

What run values do we get if we add 1 event to each play type, as we did in
Table 8-4 for the 1959 National League? Figure 8-6 displays the results. As we
expected, each walk and hit has more run value for the better offensive team, the
Cleveland Indians. Even each out is more damaging to the Indians than to the
Twins. This is because each out is one less opportunity to score runs, and since the
Indians are more productive, the out has a greater negative effect on runs scored.
As we suspected from our observation of the curves in Figure 8-4, the run values
of plays appear to vary depending on the run productivity of the team. In particu-
lar, the linear models of run production (Lindsey, Palmer, regression, and even
Total Average) appear to be special cases of this more general simulation model.
Essentially, they provide good estimates of run values for average or typical per-
formances, but do not account for changes in run value for more extreme cases.

Does this mean that they are not useful in evaluating offensive performance
of players? Not necessarily, as we shall see.

Where Do We Stand?

After reviewing many models, we find that they really divide into two groups:
additive and product models.

---

[13] The fifth highest total in this century and the highest since the Boston Red Sox posted 1027
runs in 1950.

[14] The number of innings was estimated by multiplying the number of games by 9.

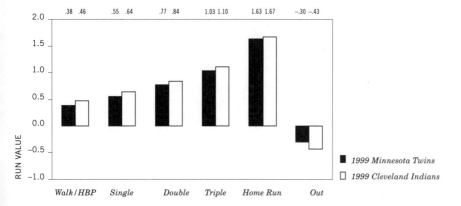

FIGURE 8-6   Run values of different plays as estimated by the DLSI model for the 1999 Cleveland Indians and Minnesota Twins.

Additive Models

These are the models in which each play is given a value. The values for all plays which occur are summed. In order to account for differences in opportunity, the sum is divided by a representative total for the number of chances to bat. The models often differ in the set of plays considered.

For Batting Average, hits are the only plays considered. Each hit has the same value, 1. These values are summed, one for each hit, and then divided by the number of at-bats, the quantity that represents opportunity. Slugging Percentage is the same as Batting Average, except that each hit is given a different value, the number of bases attained. On-Base Percentage is also the same as Batting Average, except it includes more plays, walks and hit by pitcher; it also expands the number of opportunities considered from just at-bats to (almost) all plate appearances.

Total Average, also an additive model, combined elements of the Slugging Percentage and the On-Base Percentage. Each hit was given a value equal to the number of bases, as in Slugging Percentage, and walks/HBPs were each given the value 1. Total Average also extended the plays included by giving each stolen base the value of 1. A notable innovation in Total Average was using the number of outs (not at-bats or plate appearances) as the dividing measure of opportunity. Total Average provided a significant improvement over the three MLB-recognized measures of offensive prowess (AVG, SLG, OBP) in its ability to estimate annual team run production per game.

Lindsey and Palmer basically used the same framework as in Total Average, but used estimates of the average number of runs each play produced as the play values. They developed these values from play-by-play analysis of actual (Lindsey) and simulated (Palmer) games. This gave their values a more solid logical foundation than the other additive models.

Finally, several researchers used least squares linear regression on annual team offensive data to derive play values comparable to those of Lindsey and Palmer. The play values derived from regression provided the best fit (as measured by RMSE) to annual team run production per game. However, in some cases (as we observed, for example, with the high value attributed to sacrifice flies), the values obtained from regression may have captured other attributes inappropriate for the evaluation of individual players.

Product Models

We have examined two types of product models. One, the BRA model, *multiplies* On-Base Percentage and Slugging Percentage. This is a departure from the additive model approach, where the weights from different events are simply added. Here, after weighting the events, the result of getting on base is multiplied by the weighted counts of events. Cook's Scoring Index also used this principle, as did James in his Runs Created model, which produced the best fit to team data in this group.

The second product model type is the DLSI simulation developed by D'Esopo and Lefkowitz (and generalized by Cover and Keilers). Here the rules of baseball were used to develop formulas which estimated the expected runs scored per inning if millions of simulated games of a simple form of baseball were played. The model says that the average number of runs per inning is the average number of runners to reach base minus the average left on base:

$$DLSI = \bar{R} = \bar{B} - \bar{L}$$

The average number to reach base should look familiar:

$$\bar{B} = \frac{3p}{1-p}$$

Remember that a good estimate for p is the On-Base Percentage, the fraction of plate appearances in which the team gets on base. Since not getting on base means that you were out, then $1 - p$ is the fraction of plate appearances in which the team gets out. So, $\bar{B}$ is a ratio of on-base events to out events. This ratio has elements of to On-Base Percentage (ratio of on-base events to plate appearances) and Total Average (the ratio of bases to outs).

Another interesting feature of the DLSI model is the assumption that each player to reach base is a potential run, and the team's inability to advance the runner *subtracts* from this value to find the number that actually score. At this point it diverges from Total Average, which takes the basic ability to get on base and then *adds* extra bases from hits. Another way to perform this subtraction is through multiplication by a value less than 1. This is what you do every day when you get a discount at a store. You pay less than the retail price, but instead of getting some amount off the retail price, you pay a fraction of the retail price, where the fraction is a number less than 1. So, DLSI could also be calculated as follows:

$$DLSI = \bar{R} = \bar{B} \times fraction$$

In this case, the fraction is the percentage of runners that score. So we see that the D'Esopo-Lefkowitz model can be viewed as another product model, a variant of BRA, where $\bar{B}$ assumes the role of On-Base Percentage and the discounting fraction assumes the role of the Slugging Percentage.

In order to summarize how well each model's estimates are correlated with run production, we took our most complete set of team data (1954–1999) and found the best line (and its associated RMSE) for each model for all the 46 years of data as a whole (instead of within each year, as was done in Chapter 6). The models are listed in Table 8-6—according to RMSE from low to high, so that the models at the top were correlated best with run production. We have identified each model as an additive or product model. To give some perspective, the highest team run production in this era was 6.228 runs per game (the 1999 Cleveland Indians), the lowest was 2.858 (the 1968 Chicago White Sox), and the average was 4.323. The standard deviation is 0.563 runs per game. Remember that the standard deviation is the RMSE for the simplest of models, picking the average (4.323) as the estimate for all teams in all years.

We notice right off the bat that the MLB sanctioned models (AVG, OBP, and SLG) provide the worst correlation with run production. Still, as poor as AVG, SLG, and OBP are as estimators, they do reduce the RMSE substantially, from .563 down to the .3 to .2 runs per game range. Major improvements are found by adding OBP and SLG to obtain OPS, or by multiplying them to obtain BRA. From this point on, improvements in estimation are much less dramatic, no more than a reduction of RMSE from .16 to .14 runs per game. OPS and BRA have definite advantages in the simplicity of their calculation (especially if SLG and OBP are already at hand), but the four additive and product models

Model	Abbreviation	Type	RMSE (Runs/Game)
Regression (without SF)	LSLR	Additive	.1423
Runs Created (Tech-1)	RC/G	Product	.1459
Linear Weights	LWTS/G	Additive	.1489
D'Esopo-Lefkowitz Scoring Index	DLSI	Product	.1526
Batter's Run Average	BRA	Product	.1565
Total Average	TA	Additive	.1591
Runs Created (Basic)	RC/G	Product	.1595
On-Base plus Slugging	OPS	Additive	.1595
Slugging Percentage	SLG	Additive	.2175
On-Base Percentage	OBP	Additive	.2529
Batting Average	AVG	Additive	.3169

TABLE 8-6   RMSEs for Various Models of Team Run Production per Game (1954–1999)

at the top of the list have the edge when it comes to fit (lower RMSE). DLSI and LWTS/G have additional credibility because of their construction through a logical analysis of the effect of plays within games. The regression model has the best fit, but this is really a *fait accompli,* since the model was designed from this same data.

Player Evaluations in the Best Models

Despite the differences in these models, they demonstrate a remarkable similarity in their relative evaluations of players. Let's look at the 740 players with more than 5000 plate appearances. Figure 8-7 plots the evaluation of these players by the Runs Created model versus the Linear Weights model. Each point represents the evaluation of a player's hitting career using the two models. Several players with extraordinarily high evaluations (Ruth, Williams, Gehrig, Hamilton, and Thomas) are noted. We see that both models place these players at extremely high levels. The other players form a very tight band; when LWTS/G rates a player highly, RC/G does so as well.

The line shown in Figure 8-7 is the best line fit of the player evaluations by the two models. If we examine the variation of RC/G player evaluations about this line, we find that the differences between the RC/G evaluation and the line have a standard deviation of .17 runs per game. This means that using LWTS/G, we can predict the RC/G measure to within .17 runs per game for two-thirds of the players and within .34 runs per game for 95 percent of the players. Given

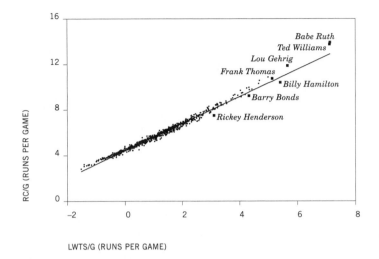

FIGURE 8-7   Career player run production estimated by the Runs Created and Linear
Weights models.

that RC/G player evaluations can range from 2 to 14 runs per game, this is very
good agreement.

There is some indication that LWTS places greater value on speed than RC
does. We have identified three players whose RC/G evaluations are low given
their LWTS/G evaluations: Billy Hamilton, Barry Bonds, and Rickey Henderson,
all exceptionally good runners. Other players with similar LWTS/G ratings have
higher RC/G ratings.[15]

One way to examine this in more detail is to create a *residual plot,* which
shows the difference between the actual value and a predicted value. Here, the
residual is the difference between the actual RC/G player evaluation and the
RC/G evaluation predicted from LWTS/G (represented by the line in Figure 8-7).
Figure 8-8 presents the residual plot for the best line in Figure 8-7.

As an example, consider Lou Gehrig. The RC/G evaluation of Gehrig's run
production is 11.84 runs per game. The LWTS/G evaluation of Gehrig's run pro-

---

[15] According to *The Hidden Game of Baseball*, Palmer's simulation produced lower SB values,
closer to .20 runs. He raised the value to .30 runs since he was persuaded that stolen bases
were more apt to occur in close games than they were to occur randomly as assumed in his
simulation. This rationale for increasing the SB value is weak, whether a game is close or not
does not enter into run production. It is possible that stolen bases occur more frequently in
base-out situations (such as runner on first base and two outs) when the extra base has more
value. Within the context of Palmer's model, this rationale for increasing SB value makes more
sense than the "close game" argument he gave in *The Hidden Game of Baseball*.

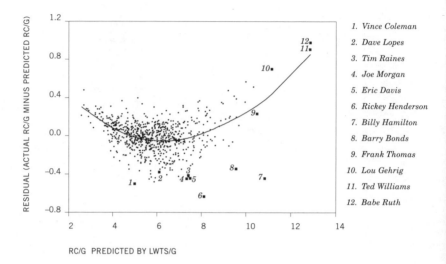

FIGURE 8-8   Career player run production: RC/G residuals versus RC/G values predicted by LWTS/G.

duction is 5.661 runs per game above average. The best line in Figure 8-7 says that RC/G and LWTS/G are so closely related that a good estimate of a player's RC/G can be found from LWTS/G using the following formula:

$$RC/G\ estimate = 4.41 + 1.1914 \times LWTS/G$$

For Gehrig, this means that:

$$RC/G\ estimate = 4.41 + 1.1914 \times 5.661 = 11.15$$

The residual for this estimate is simply the difference between the actual RC/G evaluation (11.84) and the RC/G estimate predicted by LWTS/G (11.15) or .69 runs per game. As we can see from Figure 8-8, this is a very large residual compared to most others, which lie between −.4 and +.4 runs per game. So, even though RC/G and LWTS/G both agree that Gehrig had one of the best offensive records of all players, there is some disagreement over the degree to which it was better.

What the residual plot allows us to do is to investigate the departures from the best line in Figure 8-7 in greater detail. We have noted some other players with large negative residuals—including Hamilton, Bonds, and Henderson, who have large departures from the line. Now the residual plot allows us to see other players whose RC/G values are below the best line in Figure 8-7, and several are likely to be familiar to contemporary fans: Vince Coleman, Eric Davis, Joe Morgan, Tim Raines, and Dave Lopes were all speedy players. The plot provides further evidence that LWTS and RC appear to differ over the value of speed (e.g., stolen bases) in producing runs.

The residual plot has an added interesting feature. Looking back at Figure 8-7, we might have noticed some slight curvature in the relationship between RC/G and LWTS/G. If the relationship were straight we would expect the best line to shoot right through the cloud of points, with some points below the line and others above it in all areas of the plot. However, points on the extreme left end of Figure 8-7 *and* those at the extreme right end (e.g., Ruth, Williams, Gehrig) tend to be above the best line. Does this tell us the relationship is not a straight line, but curves instead? The residual plot in Figure 8-8 confirms this. The line in that plot indicates a smooth fit that balances points above and below. This line indicates the true curved nature of the relationship, and makes it easier to see by accentuating the curvature. (Still, looking closely at Figure 8-7, one *can* see the curvature there as well.)

What is the reason for this curvature? Both models fit team run production data very well and seem to agree very well in the general player evaluations. But disagreement between the models tends to increase as we depart—in either direction—from the average players. And this seems to be especially true for the Olympian players such as Gehrig.

Player Evaluations on an Average Team

Perhaps the reason for the discrepancy between the results predicted by the RC/G and LWTS/G models lies not with the models themselves but with *how they are applied* to evaluate players. The LWTS model, when applied to player data, rates the player on how many more runs are created by the player than by an average player. In Figure 8-7, we see that it is possible for a player's LWTS rating to be negative. That is, the player produces fewer runs than the average player. Of course, in Figure 8-7, we were looking only at players who had substantial major league careers, so the great majority had to be productive offensively. Players with negative ratings (such as the infielders Mark Belanger, Ozzie Guillen, Larry Bowa, Don Kessinger, and Bobby Richardson, as well as the catchers Jim Hegan and Bob Boone) must have had very valuable defensive skills in order to compensate for their lack of run production.

All product models (such as RC) take a different approach. They evaluate the player not *relatively* (with respect to an average player) but *absolutely*—in isolation, not within any standard context. The player is evaluated in accordance with how well a team composed exclusively of that player would produce runs.

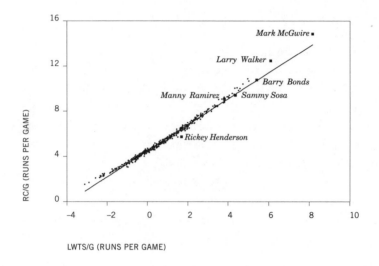

FIGURE 8-9   1998 player run production estimated by the RC and LWTS models.

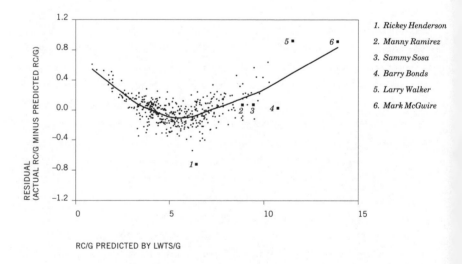

FIGURE 8-10   1998 player run production: RC/G residuals versus RC/G values predicted by LWTS/G.

Therefore, product models always produce a positive result for their evaluation of players. They estimate the cumulative number of runs produced, while LWTS estimates a differential between this player and the average player. The advantage to the RC method is that it is not necessary to find or define a standard against which to evaluate each player.

Unfortunately, there is a down side to product models as well. They tend to be unrealistic for players at either end of the offensive production spectrum. Let's look at a very extreme example in recent memory, Mark McGwire in 1998. Not only did McGwire hit home runs at a record pace in 1998, but when he wasn't trotting around the bases he very frequently walked to first base. McGwire had 162 walks (which outnumbered his 152 hits); next highest in walks was Barry Bonds, with 130. The reason, of course, for walking McGwire is to avoid his power and leave it to the next man in the lineup to knock in the runs.

But what if the next batter in the lineup is . . . Mark McGwire? The walk then becomes an extraordinarily effective force in producing runs. There is really no rationale for pitching carefully and giving the previous batter (Mark McGwire) a walk just so you can face the next batter (Mark McGwire) with an additional runner on base. This is just the situation created when the product models are used to estimate the cumulative number of runs produced by a lineup consisting of one player. The method evaluates players in a context which is outside the realm of possibility (imagine a team of McGwire clones!) and so exaggerates the effectiveness of players at the extreme ends of offensive productivity.

Figure 8-9 is a plot similar to that in Figure 8-7 except that the players evaluated are those from the 1998 season with 100 or more at bats. The figure displays the same curvature as seen in Figure 8-7. The residual plot in Figure 8-10 (constructed from Figure 8-9 just as the residual plot in Figure 8-8 was constructed from Figure 8-7) emphasizes this curvature. Mark McGwire and Larry Walker stand apart from all other players in both figures. The curvature is a result of the players being evaluated by RC/G with respect to teams composed only of that player in isolation. Is it possible to use RC/G to evaluate a player in the more appropriate context of an average team?

Let's see what happens if we analyze these players using RC/G with a different method. We will do this analysis within the context of an average team in 1998. Table 8-7 summarizes the steps in the calculation of the data used. The second column of Table 8-7 shows the average statistics for a team in 1998. These values were obtained simply by dividing the Major League totals by 30, which is the number of MLB teams. The last line of the table is the number of outs estimated from the data, calculated as follows:

$$Outs = AB - H + CS + SH + SF$$

Using this formula, we estimate that the 1998 Average Team had 4241.3 outs in the season:

$$Average\ team\ outs = 5570.6 - 1482.97 + 50.13 + 56.83 + 46.73 = 4241.3$$

Now what would happen if we replaced an average player on this team with the 1998 version of Mark McGwire (something every GM dreams of). Since the number of outs a team has in a season should be relatively fixed, we will do this by preserving the number of outs the average team had in the 1998 season. The third column lists McGwire's impressive 1998 offensive data. Using the same formula, we estimate that McGwire required 361 outs to achieve his totals:

$$McGwire\ outs = 509 - 152 + 0 + 0 + 4 = 361$$

We will now replace one average player on the 1998 Average Team with Mark McGwire. To do this, we will first calculate what the 1998 Average Team's data would look like if we subtracted one average player with the same number of outs as McGwire had in 1998. So, removing this one average player reduces the Average Team's data by a percentage equal to McGwire's outs divided by the Average Team's outs, or:

$$361 / 4241.3 = 8.5\%$$

The fourth column of Table 8-7 shows the data for the 1998 Average Team with one less average player. Each value is simply 100 percent − 8.5 percent = 91.5 percent of the 1998 Average Team values in the second column of the table.

Now, in order to see what the 1998 Average Team would have been like with Big Mac replacing one of its average players, all we have to do is add Mac's data in column 3 to the reduced team data in column 4. The resulting data are shown in the fifth column of Table 8-7. Notice that the number of outs is exactly the same as that for the 1998 Average Team in the second column of Table 8-7.

Out of curiosity, we might want to see how much this team differs from the 1998 Average Team. These results are displayed in the sixth column. They were calculated by subtracting the second column from the fifth column. The Average Team with Mac would have had about 35 more at-bats and 26 more hits. This is because McGwire had a better-than-average chance of getting a hit, so an equal number of outs produces more hits and thus more at-bats. With Mac, the Average Team would have had a whopping 56 more home runs and 115 more walks,

Play	Avg. Team	McGwire	Team Without 1 Avg. Player	Avg. Team With Mac	Avg. Team With Mac Minus Avg. Team
AB	5570.6	509	5096.5	5605.5	34.9
H	1483.0	152	1356.7	1508.7	25.8
2B	291.3	21	266.5	287.5	–3.8
3B	30.0	0	27.4	27.4	–2.6
HR	168.8	70	154.4	224.4	55.6
BB	548.2	162	501.6	663.6	115.3
HBP	52.9	6	48.4	54.4	1.5
IBB	35.6	28	32.5	60.5	25.0
SB	109.5	1	100.1	101.1	–8.3
CS	50.1	0	45.9	45.9	–4.3
SH	56.8	0	52.0	52.0	–4.8
SF	46.7	4	42.8	46.8	0.0
Outs	4241.3	361	3880.3	4241.3	0
RUNS CREATED	•	•	770.44	950.48	180.04

TABLE 8-7   Offensive Data for Average 1998 Team If Mark McGwire Replaced an Average Player

approximately 33 percent and 20 percent increases, respectively. Incredibly, the number of intentional walks would increase by about 70 percent.

We are now ready for the final steps. First, we apply the RC formula to the data for the 1998 Average Team without one average player (fourth column of Table 8-7):

$$RC = \frac{(H + BB + HBP - CS - GIDP)\,[TB + .26\,(BB - IBB + HBP) + .52\,(SH + SF + SB)]}{AB + BB + HBP + SH + SF}$$

Doing this, we find that this team is expected to generate a total of 770.44 runs using 3880.3 outs. The next step is to apply the RC formula again, this time to the data for the 1998 Average Team with McGwire (fifth column of Table 8-7). We find that this team is expected to generate a total of 950.48 runs using 4241.3 outs.

So we conclude that when his performance is considered within the context of an average team in the 1998 season, McGwire would add $950.48 - 770.44 = 180.04$ runs using 361 outs, the difference between the runs the team would be expected to score with and without him. Since 361 outs is equivalent to $361/27 = 13.37$ games, McGwire's contribution is $180.04/13.37 = 13.47$ runs per game. This is 1.36 runs per game less than the 14.83 value estimated by James' standard method for evaluating individual players.[16] So, placing McGwire within a realistic team context reduces his RC/G estimate by almost 10 percent.

To what extent are other players affected by making a similar adjustment? Figure 8-11 plots the change in Runs Created per Game for each player versus the original Runs Created per Game estimate. What has happened is that the adjusted RC evaluation method has reduced the RC/G estimates for the best and worst players while providing little or no effect on those more typical players in the center of the spectrum.

What are the implications of this adjustment relative to the comparability of the RC and LWTS evaluations of players? Figure 8-12 replicates our analysis from Figure 8-9, except that that Figure 8-12 uses the adjusted RC player evaluation method instead of James' standard method. A new best line is plotted, and it now appears to go straight through the points, with no bend in points above the line at the extreme ends. This is confirmed when we examine the adjusted RC/G residuals in Figure 8-13. Not only is the plot flat, with no evident curvature, but the spread of the points has been reduced as well.

This analysis indicates that much of the difference between the Linear Weights and Runs Created models lies not only in the models themselves, but

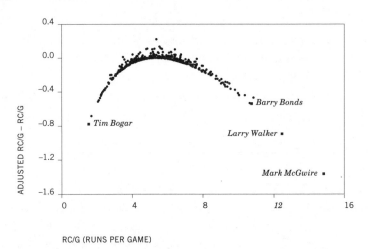

FIGURE 8-11   Change in Runs Created evaluation of 1998 players:(adjusted RC/G – standard RC/G) versus standard RC/G.

---

[16] Apparently, James has developed a similar method for adjusting his Runs Created evaluation for individual players. We have not seen the method, but the description in the *1999 Big Bad Baseball Book* indicates that it follows principles similar to those presented here, except that at-bats rather than outs are used as the basis for player replacement.

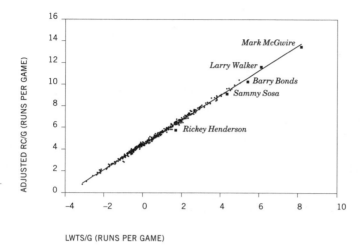

FIGURE 8-12   1998 player run production using the adjusted RC and LWTS models.

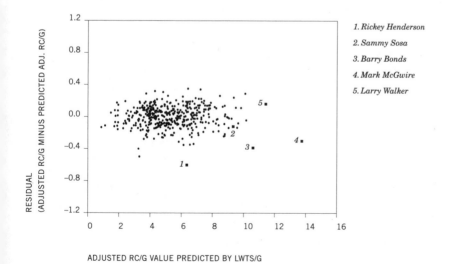

FIGURE 8-13   1998 player run production: adjusted RC/G residuals versus adjusted RC/G values predicted by LWTS/G.

how they are used to evaluate players. Also we see that each model has its advantages. If we wish to evaluate players, the Linear Weights model is simpler than the Runs Created model. However, the Runs Created model has greater flexibility in its ability to analyze run production beyond the context of an average team. And yet even with these differences, there is a strong correlation between:

1. each model with team runs scored; and

2. the two models themselves.

Thus, in evaluating players, a fan can't go too far wrong using either of these models for evaluating players.

Sorting Out Strengths and Weaknesses

Let's review what we have learned about evaluating players:

1. The batting performance of players should be measured with respect to the number of runs they contribute to team offense.

2. If these player evaluations are normalized with respect to the number of opportunities the player had to produce offensively, the number of outs is a better measure than at-bats or plate appearances for this purpose. Number of outs can be converted to equivalent games by dividing by 27 (a theoretical standard).

3. The standard measures used by Major League Baseball and the media were the worst evaluators of offensive performance among those reviewed.

4. A model's correlation with runs produced by teams in a season is an important measure to establish its capability to estimate run production. However, to blindly use one model over another just because its correlation is higher (or its predictive error is lower) is not wise. The model should be checked to insure that its structure is a logical representation of our understanding of baseball.

5. A simulation model (the DLSI model) emulates the most basic elements of baseball play in such a way that run production can be reasonably modeled *without* the need to actually play out the simulation using random number generators.

6. Models can be described as either additive or product. Product models are a better reflection of the curved nature of run production, but additive models are simpler to use in evaluating players.

7. The best of the additive models (Linear Weights) and product models (Runs Created) are related to logical constructs derived from actual baseball play. Linear Weights are constructed from data analyses and simulation, while Runs Created is related to the basic equations of the D'Esopo-Lefkowitz model. These two models are strongly correlated in their player evaluations, especially after the Runs Created model is applied in the context of an Average Team.

There are many issues we have left unresolved. Much of this book places great emphasis on the difference between observed performance (e.g., a player's batting average in a season) and ability (e.g., the underlying probability of getting a hit). These past three chapters have focused on reducing the standard multidimensional array of *observed* player offensive data into a single value that is strongly correlated with runs scored. We have not discussed the relation of this value to some underlying parameter for a player's ability to generate runs. Actually, it is not too difficult to calculate confidence intervals for such a parameter for many of these measures.

We have also skirted the issue of adjusting player evaluations for different playing conditions. This issue as it relates to comparing players from different eras is a cottage industry in itself, with many worthy publications that address the topic. Interested readers may wish to examine books such as Michael Schell's *Baseball's All-Time Best Hitters* to see how this question has been addressed.

However, if we rank players within each decade as we did for many measures in Chapter 6, we are in essence making a gross adjustment for the nature of the game in each decade. Given the capabilities demonstrated by the Runs Created model, the list of outstanding hitters in Table 6-14 is a very reasonable compilation of the 36 greatest career performances in generating runs. As great as these players were, many were dogged by the question of whether they produced in clutch situations. (As followers of the Phillies, we well remember fellow Philadelphians' doubts about Mike Schmidt in this regard—that is, until he led the Phillies to a World Championship in 1980.) In the next chapter, we will examine the clutch hitting issue and put some of the work in Chapter 7 to use in attempting to quantify contributions to winning.

MEASURING
CLUTCH PLAY

Chapter 9

"At its most critical moments, baseball chooses its heroes and goats with the randomness of a carnival barker's rickety spinning wheel. Where she stops nobody knows." So wrote Tom Verducci in his *Sports Illustrated* article describing the sporting event he would most like to have witnessed.[1] The event was Cookie Lavagetto's double in the fourth game of the 1947 World Series. The hit, with two outs in the bottom of the ninth inning, not only won the game for Lavagetto's Dodgers over the Yankees, to tie the series—it also spoiled Yankee Bill Bevens' bid to pitch the first no-hit game in World Series history.

Out of the 25 greatest moments in baseball history (as selected in the October 18, 1999, issue of *The Sporting News*), 10 (or 40 percent)[2] involve clutch hits. Here are those 10 great clutch moments, in order as they appeared in the list:

#1: Bobby Thomson's home run to win the 1951 National League pennant playoff for the New York Giants.

#2: Bill Mazeroski's home run to win the 1960 World Series for the Pittsburgh Pirates.

#4: Carlton Fisk's home run to guarantee a Game 7 for the Boston Red Sox in the 1975 World Series.

---

[1] *Sports Illustrated*, November 29, 1999, p. 80.
[2] #10 on *The Sporting News* list was St. Louis Cardinal Enos Slaughter's dash from first to score the series-winning run in the 1946 World Series. The percentage would rise to 44 percent if we include the hit by Harry Walker that initiated the play.

#6:   Kirk Gibson's two-out, ninth-inning home run, which turned defeat into victory for the Los Angeles Dodgers in Game 1 of the 1988 World Series.

#14:   Bucky Dent's home run to win the 1978 American League East pennant playoff for the New York Yankees.

#16:   Joe Carter's home run to win the 1993 World Series for the Toronto Blue Jays.

#21:   Chris Chambliss's home run to win the 1976 American League pennant for the New York Yankees.

#22:   George Brett's home run to win Game 3 of the 1980 American League Championship Series for the Kansas City Royals.

#24:   Dave Henderson's two-out, ninth-inning home run in Game 5 of the 1986 American League Championship to keep the Boston Red Sox alive.

#25:   Cookie Lavagetto's double (described above).

Another 3 events (12 percent[3]) are moments in which hitters were *unsuccessful* in clutch situations:

#9:   Vic Wertz's long drive to center field, caught by Willie Mays, in the eighth inning of a tied Game 1 of the 1954 World Series.

#13:   Willie McCovey's line drive into the final out of the 1962 World Series, with the series-winning runs in scoring position.

#18:   Yogi Berra's bid for an extra-base hit with two runners on which was caught by Sandy Amoros in the sixth inning of Game 7 of the 1955 World Series.

Most baseball fans are familiar with at least some of these moments. While we all may have favorites that were not included and may disagree with the rankings, by and large, these are indeed great moments in the history of the sport.[4] But what exactly is it that makes them so?

All thirteen occurred in important games. Eight occurred in World Series games. Of the remaining five, a league championship was at stake in four and a

---

[3] Two other moments involved plays in which the hitter was unsuccessful, but an error by a fielder produced a positive result for the team at bat. *The Sporting News*'s #8 was Bill Buckner's error in Game 6 of the 1986 World Series; and #23 was Catcher Mickey Owen's dropped third strike, which would have been the final out of Game 4 in the 1941 World Series.

[4] We are partial to events involved in deciding a pennant race, including two on the final day of the 1950 National League season. Phillie Richie Ashburn threw out Cal Abrams' potential winning run in the bottom of the ninth inning to preserve the tie game between the two pennant contenders, Philadelphia and Brooklyn. In extra innings, Dick Sisler's home run won the pennant for the Phillies. (Not so incidentally, the climax of dramatizations like *The Natural*, *Damn Yankees*, and *Major League* is a game that decides the pennant, not the World Series.)

division championship in another. Of course, it isn't just the game that makes these moments special—it is the situation within the game. Eight of these moments were the final play of a game in which victory was decided. Four of them (the Thomson, Gibson, Carter, and Lavagetto events) turned defeat into victory; three (Mazeroski, Fisk, and Chambliss) provided the winning edge in a tie game; and one (McCovey) was the last failed attempt to attain victory.

Of course, few games have the inherent drama of a pennant clincher or the seventh game of the World Series. But throughout the season, even in inconsequential games between two also-ran teams, spectators are gripped by dramatic situations in which the game hangs in the balance, where one play, even just one ball or strike, may be the difference between victory and defeat. In trying to define which of these should be called a "clutch situation" (an admittedly vague term), we've come up with these criteria:

- *The score is close.* The fan must feel that whatever happens in the next play will provide a decisive edge in the score and thereby determine the outcome of the game.

- *The situation is late in the game.* In the early innings, there is always the feeling that there will be time to come back. But as the game proceeds and the shadows lengthen (in those increasingly rare games played outdoors and in the daytime), each play gains in importance, especially when the score is close.

- *Runners are on base.* It is possible (as we will see later) for situations with no runners on to be important, but the drama increases when runners reach base and the probability of scoring increases.

- *Two outs is more dramatic than one out, which is more dramatic than no outs.* As outs increase, the number of opportunities to score in the inning decreases, placing greater value on the opportunities remaining.

Clutch Hits

If we pick a player and say that he hits in the clutch, what exactly do we mean? Do we mean that he hits better when runners are on base? When the score is close? When runners are on and there are two outs?

To gain some perspective, we will examine how well Major League Baseball players as a group have hit in various situations. From 1985 through 1993, the

Elias Sports Bureau published the annual *Elias Baseball Analyst*. The books provided hitting data for each league in the following game situations:

- Entire season—all game situations.
- Leading off an inning.
- Runners on—at least one runner on base.
- Runners in scoring position—at least one runner on second or third base.
- Runners on and two outs.
- Runners in scoring position and two outs.

- *Late Inning Pressure (LIP)*—a situation in the seventh inning or later with the score tied or with the batter's team trailing by one, two, or three runs (or four runs and the bases loaded). The definition (and name) of this situation was the creation of the Elias Sports Bureau, who wanted to capture those circumstances in which the game was close and there was a decreasing opportunity to change the outcome. LIP has become standard enough to be listed in *The Dickson Baseball Dictionary*.

- Leading off an inning in a LIP situation.
- Runners on in a LIP situation.
- Runners in scoring position in a LIP situation.

Table 9-1 provides American League total at-bats, hits, and overall batting averages for the years 1984, 1986–88, 1990, and 1992. (We restricted the com-

Situation	AB	H	AVG
All	464,057	121,269	.261
Leading Off	111,756	28,856	.258
Runners On	201,183	54,029	.269
Runners/Scoring Position	114,773	30,144	.263
Runners On/2 Out	84,996	21,208	.250
Scoring Position/2 Out	54,476	13,187	.242
Late Inning Pressure (LIP)	66,824	16,901	.253
LIP Leading Off	16,745	4,156	.248
LIP Runners On	28,447	7,373	.259
LIP Runners/Scoring Position	16,165	4,048	.250

TABLE 9-1　　Batting Averages in Various Game Situations, American League, 1984, 1986–88, 1990, 1992

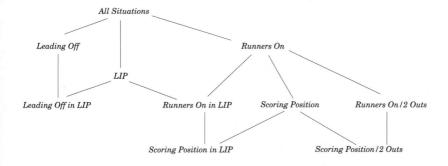

FIGURE 9-1   Schematic depiction of situational subsets (each line connects a subset at a lower level to a superset at a higher level).

parison to the American League to exclude the batting of pitchers and so preserve consistency in the types of hitters analyzed.)

Before comparing averages in the different situations shown in Table 9-1, we must understand that many of these situations are not independent of one another. Let's consider "Runners On" and "Scoring Position." When a runner is in scoring position, there are always runners on base. This may be obvious, but it is important to understanding the data in Table 9-1. Every at-bat tabulated in the Runners in Scoring Position row of Table 9-1 is also included in the totals for the Runners On row; one situation is a subset of the other. So, the two rows cannot be compared directly. The Runners in Scoring Position row must be subtracted from the Runners On row to create Runners *Not* in Scoring Position. The AVG in this situation can then be compared to the Runners in Scoring Position AVG to determine the effect of having runners in scoring position as opposed to having a runner only on first base.

Figure 9-1 shows the dependencies between the situations in Table 9-1. Each line connects one situation to another situation (at a higher level in the hierarchy) of which it is a subset. For example, Scoring Position is connected to Runners On, which is at a higher level. In a general sense, the lower the situation is in Figure 9-1, the more we regard it as a clutch situation.

Each line in Figure 9-1 represents a possible comparison of situations:

- Leading Off an Inning versus Not Leading Off an Inning (Overall or in LIP).

- Runners in Scoring Position versus Bases Empty (Overall or in LIP).

	OVERALL			LATE INNING PRESSURE		
	AB	H	AVG	AB	H	AVG
Leading Off	111,756	28,856	.2582	16,745	4156	.2482
Not Leading Off	352,301	92,413	.2623	50,079	12,745	.2545
Difference	•	•	−.0041	•	•	−.0063

TABLE 9-2   Leading Off an Inning versus Not Leading Off, American League, 1984, 1986–88, 1990, 1992

	OVERALL			LATE INNING PRESSURE		
	AB	H	AVG	AB	H	AVG
Scoring Position	114,773	30,144	.2626	16,165	4048	.2504
Bases Empty	262,874	67,240	.2558	38,377	9528	.2483
Difference	•	•	.0069	•	•	.0021

TABLE 9-3   Runners in Scoring Position versus Bases Empty, American League, 1984, 1986–88, 1990, 1992

	OVERALL			LATE INNING PRESSURE		
	AB	H	AVG	AB	H	AVG
Scoring Position	114,773	30,144	.2626	16,165	4048	.2504
First Base Only	86,410	23,885	.2764	12,282	3325	.2707
Difference	•	•	−.0138	•	•	−.0203

	TWO OUTS		
	AB	H	AVG
Scoring Position	54,476	13,187	.2421
First Base Only	30,520	8021	.2628
Difference	•	•	−.0207

TABLE 9-4   Runner in Scoring Position versus Runner on First Base Only, American League, 1984, 1986–88, 1990, 1992

- Runners in Scoring Position versus Runner on First Base Only (Overall, With Two Outs, or in LIP).

- Two Outs versus None/One Out (With Runners On or With Runners in Scoring Position).

- LIP versus No LIP (Overall, When Leading Off, or With Runners in Scoring Position).

Let's look at each comparison separately, as presented in Tables 9-2 through 9-6.

Leading Off an Inning vs. Not Leading Off

Hitters appear to bat about .004 points less when leading off than when not leading off. The difference is not appreciably changed in Late Inning Pressure situations.

Runners in Scoring Position vs. Bases Empty

Batters hit better with runners in scoring position than when bases are empty, showing an increase in AVG of .007. However, the effect is not significant in LIP situations. This effect may be a reflection of pitching skill, which produces the scoring situation. Since less capable pitchers are likely to put runners in scoring position more often than better pitchers do, the increase in AVG may result from the greater chance that the pitcher in scoring-position situations is on average worse than one encountered in bases-empty situations. It also seems reasonable that the effect would be less in LIP situations, when the manager is more likely to replace the less effective pitcher with a more capable reliever.

Runner in Scoring Position vs. Runner on First Base Only

In this scenario we are considering only situations where runners are on base. What we are examining is how batters perform when a runner is on first base only compared with all situations with a runner in scoring position. We see large effects here: overall, batting average is reduced by .014. In LIP situations and with two outs, batting averages are even lower, reduced by about .020 with runners in scoring position, as compared to when a lone runner is on first base. This difference may result in part from the first baseman playing near the base in order to hold the runner on first, leaving a hole between second and first. The batter is more likely to have this hole when there is a runner on first base only; if there are runners in scoring position, the first baseman is less likely to play near the first base bag.

	RUNNERS ON			FIRST BASE ONLY		
	AB	*H*	*AVG*	*AB*	*H*	*AVG*
Two Outs	84,996	21,208	.2495	30,520	8021	.2628
0 or 1 Out	116,187	32,821	.2825	55,890	15,864	.2838
Difference	•	•	−.0330	•	•	−.0210

	SCORING POSITION		
	AB	*H*	*AVG*
Two Outs	54,476	13,187	.2421
0 or 1 Out	60,297	16,957	.2812
Difference	•	•	−.0392

TABLE 9-5   Two Outs versus None/One Out, American League, 1984, 1986–88, 1990, 1992

	OVERALL			LEADING OFF		
	AB	*H*	*AVG*	*AB*	*H*	*AVG*
LIP	66,824	16,901	.2529	16,745	4156	.2482
No LIP	397,233	104,368	.2627	95,011	24,700	.2600
Difference	•	•	−.0098	•	•	−.0118

	SCORING POSITION		
	AB	*H*	*AVG*
LIP	16,165	4048	.2504
No LIP	98,608	26,096	.2646
Difference	•	•	−.0142

TABLE 9-6   Late Inning Pressure versus No Late Inning Pressure, American League, 1984, 1986–88, 1990, 1992

Two Outs vs. None/One Out

With a runner on base, batters have a harder time getting a hit when there are
two outs as compared to when there are one or no outs. When there is only a runner on first base, batting averages drop about .021 when there are two outs.
There is an even larger drop-off, .039, when a runner is in scoring position. A
possible reason for this difference may be that better hitters are not given much
opportunity to hit with runners in scoring position and two outs; they are walked
instead.[5] Overall, the decrease is about .033.

Late Inning Pressure vs. No Late Inning Pressure

Batting averages drop in Late Inning Pressure situations. In general, the LIP
effect is a drop of about .010 in AVG. While it may appear that the effect varies
from AVG drops of .010 to .014 (for runners in scoring position), this difference is
not significant.

So we see that the game situation can have an effect on how well a batter hits.
This is likely due to the pitcher, who may draw from his reserve of strength to
make a special effort to bear down as the occasion demands. The effect can also
be the result of managerial pitching strategy: in clutch situations, he may call on
relief pitchers, who in general have more success, at least in the short run, in
getting batters out.

A Player in a Short Series

What happens if we look at the performance of an individual player in a short
series? Let's pick one of the players from the 25 greatest moments. The most
recent is Joe Carter, whose home run in the 1993 World Series is remembered by
many fans.[6] The question is, do we recall how well Carter did in his 28 total plate
appearances in the series? He had 3 sacrifice flies, no walks, no sacrifice hits,
and was not hit by a pitcher, so his total number of at-bats was 25. His 7 hits
(including 1 double and 2 home runs) gave him a .280 AVG and a .560 SLG. Since
he had no walks, and sacrifice flies are included among opportunities for getting

---

[5] Data from *The 1993 Elias Baseball Analyst* provides some evidence substantiating this
conjecture. In the American League in 1992, with two outs, a walk was almost twice as likely to
occur in a plate appearance with runners in scoring position than with a runner on first base
only.

[6] Some with more pleasure than others. One of us told his son (also a Phillies fan) that he should
be prepared to see this event in replay for the rest of his life. Coincidentally, he was the same
age (14) his father was when the Phillies collapsed in the 1964 National League pennant race.

	AB	H	2B	3B	HR	SF	AVG	SLG	OBP
Leading Off	9	2	1	0	0	0	.222	.333	.222
Runners On	11	3	0	0	2	3	.273	.818	.214
Scoring Position	5	2	0	0	1	3	.400	1.000	.250
Runners On/2 Out	2	0	0	0	0	0	.000	.000	.000
Scoring Position/2 Out	0	0	0	0	0	0	.000	.000	.000
LIP	5	1	0	0	1	0	.200	.800	.200
LIP Leading Off	3	0	0	0	0	0	.000	.000	.000
LIP Runners On	2	1	0	0	1	0	.500	2.000	.500
LIP Scoring Position	2	1	0	0	1	0	.500	2.000	.500
Overall	25	7	1	0	2	3	.280	.560	.250

TABLE 9-7   Summary of Joe Carter's Plate Appearances by Game Situation in the 1993 World Series

on base, Carter's on-base percentage (7/28 = .250) was actually *lower* than his batting average. Carter scored 6 runs and had 8 RBI.

Table 9-7 provides a situational summary of Carter's plate appearances in the 1993 World Series along the lines of those in the *Elias Baseball Analyst*. Carter led off a relatively large number of times, in about one-sixth of Toronto's innings. He performed poorly in this situation. He did very well with runners on base, and especially with runners in scoring position. However, he was not able to come through with runners on when there were two outs. In Late Inning Pressure situations, he was only successful in one at-bat. Of course, this was his legendary home run, which produced an impressive slugging percentage in his limited number of LIP situations.

So how are we to judge Joe Carter as a clutch hitter in the 1993 World Series? He was a poor lead-off hitter in many at-bats (for the simple reason that he did not often get on base—the leadoff man's main job). On the other hand, he hit well with runners on, but not with two outs. And of course, he came through *big* under Late Inning Pressure, but only once. By what criterion can we gauge his clutch performance when there are so many metrics to choose from? One solution is to reduce these multiple criteria into a single metric that "weights" his performance in different situations. None of the models we have considered so far in earlier chapters perform this kind of weighting. They all consider each event generically, without regard to the situation in which it occurred.

Is it possible to integrate a clutch effect (the effect of game situation) into player evaluation? Actually, George Lindsey's data on run production allows us to do this relatively easily.

Situation Evaluation of Run Production

Recall from Chapter 7 that Lindsey developed a table (Table 7-4) of expected runs produced in each situation (outs and runners on base) within an inning of a game. We used this table to estimate the value of each type of hit in terms of runs. These values (presented in Table 7-7) are the average or expected number of runs for each hit after the frequencies of all game situations have been considered. But different players may come to bat in these situations with frequencies different from the typical values in Table 7-4.

If we knew the situation for each plate appearance for a player, we could calculate the expected number of runs produced for the player's specific opportunities, instead of using the generic frequencies in Lindsey's calculation. In this way we would capture the batter's specific level of opportunity as well as his response to it (the results of his plate appearances).

Of course, this requires much more data than the summaries of at-bats, hits, home runs, etc. We need to know, *for every plate appearance*, the situation when the player came to bat and the situation after he came to bat. Data on each plate appearance is becoming more accessible as the years go by, but it is still relatively difficult to find in electronic form.[7] (Of course, they are available in written form in the official scoresheets, but the sheer volume of these records make them difficult to use except in very limited quantities.)

Joe Carter's batting performance in the 1993 World Series provides us with a reasonable amount of data and allows us to include a top-25 moment as well. Table 9-8 lists each plate appearance in chronological order in the series. The first column identifies the result of Carter's at-bat in standard terms. Most of the abbreviations should be readily identifiable, with the exception of FO and GO, which indicate outs caught on the fly and on the ground, respectively. The second column indicates the game in the series. The third column lists the innings, with V and H identifying the visitor and home half of each; an L in parentheses is appended to the inning if Carter was the leadoff batter. The next two sets of four columns each describe the state of the game before and after the play; the state

[7] See the Retrosheet Web Page at www.retrosheet.org for advances on this front.

			BEFORE PLAY				AFTER PLAY				
Play	Game	Inning	VRuns	HRuns	Bases	Outs	VRuns	HRuns	Bases	Outs	
■ 1B	1	2H(L)	2	0	0	0	2	0	1	0 ■	
▲ SF	1	3H	3	2	3	1	3	3	0	2 ▲	
GO	1	6H(L)	4	4	0	0	4	4	0	1	
K	1	7H	4	8	2	1	4	8	2	2	
FO	2	1H	0	0	1	2	0	0	0	3	
■ ▲ HR	2	4H	5	0	1	0	5	2	0	0 ▲ ■	
FO	2	6H(L)	5	2	0	0	5	2	0	1	
K	2	8H	6	3	2	0	6	3	2	1	
▲ SF	3	1V	2	0	3	0	3	0	0	1 ▲	
■ 1B	3	3V	4	0	0	2	4	0	1	2 ■	
FO	3	5V	4	0	0	2	4	0	0	3	
K	3	7V	6	1	13	0	6	1	13	1	
GO	3	9V	8	2	1	0	8	2	1	1	
■ 1B	4	1V	0	0	12	1	0	0	123	1 ■	
FO	4	3V(L)	3	6	0	0	3	6	0	1	
FO	4	4V(L)	7	6	0	0	7	6	0	1	
FO	4	6V	8	12	1	0	8	12	1	1	
■ 1B	4	8V	9	14	0	1	9	14	1	1 ■	
■ 2B	4	9V(L)	15	14	0	0	15	14	2	0 ■	
FO	5	1V	0	0	1	2	0	0	0	3	
K	5	4V	0	2	1	1	0	2	1	2	
GO	5	7V(L)	0	2	0	0	0	2	0	1	
FO	5	9V(L)	0	2	0	0	0	2	0	1	
▲ SF	6	1H	0	1	3	1	0	2	0	2 ▲	
GO	6	3H	0	3	0	1	0	3	0	2	
FO	6	5H	1	5	0	1	1	5	0	2	
FO	6	8H(L)	6	5	0	0	6	5	0	1	
■ ▲ HR	6	9H	6	5	12	1	6	8	0	3 ▲ ■	

TABLE 9-8   Joe Carter's Plate Appearances in the 1993 World Series (Squares Mark On-Base Events, Triangles Mark Plays That Scored Runs)

is described by the score (visiting-team runs are VRuns, home-team runs HRuns), the bases occupied, and the number of outs. Note that for Carter's Game-7 home run we have listed the number of outs after the play as three; this is done since Carter's HR ended the game, consequently ending the possibility of scoring any more runs. Plate appearances in which Carter got on base are marked by squares, and those in which he drove in a run are marked by triangles. We see that Carter got a hit in every game except Game 5, when Curt Schilling shut down Toronto completely. We also see that Carter produced runs only with the long ball—either a home run or a sacrifice fly.

Now we will reduce Table 9-8 to its essentials for our calculation. Table 9-9 shows the situation (outs and runners on base) before Carter's plate appearance, the situation after Carter's plate appearance, and the number of runs scored on the play. The three Expected Runs columns show the number of runs Toronto would be expected to score in the Before situation, the number expected to score *after* the play (including the runs which actually scored), and the Change (the difference between the Before and After situations in terms of runs).

As an example, consider Carter's first RBI of the series, in the second row of Table 9-9. As Carter came to bat, Toronto had a runner on third base with 1 out. According to Lindsey's data in Table 7-4, a team on average would score .980 runs in this situation. This is the value of the Before situation in terms of runs. After Carter's sacrifice fly, Toronto had bases empty and 2 outs. According to Lindsey's data in Table 7-4, a team on average would score .102 runs in this situation. Since the SF scored the runner from third, the expected number of runs scored after the play is $1 + .102 = 1.102$. So here a SF was worth $1.102 - .980 = .122$ runs, the change in expected runs between the Before and After situations.

A special case for this calculation occurs in the bottom half of the ninth and any subsequent innings. The expected values for runs scored in Table 7-4 are predicated on the team having 3 outs to complete the inning. However, in the bottom of the ninth, the game ends as soon as the home team scores enough runs to win the game, thus limiting the run-production capability of the home team.

For example, consider the situation in Joe Carter's final at-bat. If this were an inning in the middle of the game, Table 7-4 indicates that the Blue Jays would be expected to score .939 runs. This expectation is derived from the probabilities of scoring different numbers of runs. Toronto has a 57.1-percent chance of scoring no runs, a 16.3-percent chance of scoring exactly 1 run, an 11.9-percent chance of scoring exactly 2 runs, and a 14.7-percent chance of scoring 3 or more runs. However, Carter's situation occurred in the ninth inning, with his team trailing by a single run. So, as soon as Toronto scores two runs, the inning ends

	Pitcher	BEFORE PLAY			AFTER PLAY			Play	EXPECTED RUNS			
		Bases	Outs	Runs	Bases	Outs			Before	After	Change	
■	Schilling	0	0	0	1	0		1B	.461	.813	.352	■
▲	Schilling	3	1	1	0	2		SF	.980	1.102	.122	▲
	Schilling	0	0	0	0	1		GO	.461	.243	−.218	
	Andersen	2	1	0	2	2		K	.671	.297	−.374	
	Mulholland	1	2	0	0	3		FO	.219	0	−.219	
■ ▲	Mulholland	1	0	2	0	0		HR	.813	2.461	1.648	▲ ■
	Mulholland	0	0	0	0	1		FO	.461	.243	−.218	
	Mason	2	0	0	2	1		K	1.194	.671	−.523	
▲	Jackson	3	0	1	0	1		SF	1.390	1.243	−.147	▲
■	Jackson	0	2	0	1	2		1B	.102	.219	.117	■
	Jackson	0	2	0	0	3		FO	.102	0	−.102	
	Rivera	13	0	0	13	1		K	1.940	1.115	−.825	
	Andersen	1	0	0	1	1		GO	.813	.498	−.315	
■	Greene	12	1	0	123	1		1B	.939	1.642	.703	■
	Greene	0	0	0	0	1		FO	.461	.243	−.218	
	Mason	0	0	0	0	1		FO	.461	.243	−.218	
	West	1	0	0	1	1		FO	.813	.498	−.315	
■	Andersen	0	1	0	1	1		1B	.243	.498	.255	■
■	Thigpen	0	0	0	2	0		2B	.461	1.194	.733	■
	Schilling	1	2	0	0	3		FO	.219	0	−.219	
	Schilling	1	1	0	1	2		K	.498	.219	−.279	
	Schilling	0	0	0	0	1		GO	.461	.243	−.218	
	Schilling	0	0	0	0	1		FO	.461	.243	−.218	
▲	Mulholland	3	1	1	0	2		SF	.980	1.102	.122	▲
	Mulholland	0	1	0	0	2		GO	.243	.102	−.141	
	Mulholland	0	1	0	0	2		FO	.243	.102	−.141	
	Mason	0	0	0	0	1		FO	.461	.243	−.218	
■ ▲	Williams	12	1	3	0	3		HR	.695	3.000	2.305	▲ ■

Sum of Change = 1.231

TABLE 9-9   Change in Expected Runs for Joe Carter Plate Appearances in the 1993 World Series (Squares Mark On-Base Events, Triangles Mark Plays That Scored Runs)

and there are no more opportunities to score additional runs. This has the effect of truncating the distribution of runs to no more than 2, the number needed to win the game. This means that all of the situations in which more than 2 runs could have scored in a full inning are reduced to occurrences of 2 runs. So, in this score-inning situation, the probability of scoring two runs is 26.6 percent, the sum of the probability of scoring two runs (11.9 percent) and the probability of scoring 3 or more runs (14.7 percent). The expected number of runs scored with runners on first and second with 1 out (when trailing by 1 run in the bottom of the ninth inning) can now be estimated as follows:

$$(0 \times .571) + (1 \times .163) + (2 \times .266) = .695 \text{ runs}$$

This result is .244 runs less than in other innings because of the limitation of not being permitted to score more runs than the number needed to win.[8]

Performing these calculations for every play, we find the value of each plate appearance for Carter in terms of runs. This is similar to the calculation we did in Chapter 7 to determine the average value of each type of hit in terms of runs. There we generated a different After situation for each possible Before situation, then calculated the Change in expected runs. Here we are taking an actual record of plate-appearance results and finding their values in terms of expected runs produced. If we total the run values in the Change column of Table 9-9, we find that the net result of Carter's batting in the 1993 World Series was +1.231 runs. This means that Carter produced 1.231 more runs (.205 more runs per game) than an average batter would be expected to produce in the same set of Before situations.

What we have done is integrate clutch effects into the estimate of run production for a specific batting performance. We have replaced the various categories of clutch situations of Table 9-7 with a single value of run production weighted by the situation in which each event occurred.

To see that we have actually accomplished this, let's examine an alternate- or parallel-universe batting performance for Joe Carter. The alternate performance in Table 9-10 is one that Phillies' fans wish had actually occurred. Focus your

[8] This is an underestimate of the actual value because of baseball's ruling that all runs that score on a game-winning home run are counted. There is a chance of scoring 3, 4, or 5 more runs when trailing by 1 run in the bottom of the ninth inning. (Carter's HR, which resulted in 3 runs scored, is an example of this.) However, since these scores can only be achieved with a final HR, they are much less probable than in normal circumstances. For simplicity (and the lack of data), we use the lower value here.

	Pitcher	BEFORE PLAY		AFTER PLAY				EXPECTED RUNS			
		Bases	Outs	Runs	Bases	Outs	Play	Before	After	Change	
■	Schilling	0	0	0	1	0	1B	.461	.813	.352	■
▲	Schilling	3	1	1	0	2	SF	.980	1.102	.122	▲
	Schilling	0	0	0	0	1	GO	.461	.243	−.218	
	Andersen	2	1	0	2	2	K	.671	.297	−.374	
	Mulholland	1	2	0	0	3	FO	.219	0	−.219	
■ ▲	Mulholland	1	0	2	0	0	HR	.813	2.461	1.648	▲ ■
	Mulholland	0	0	0	0	1	FO	.461	.243	−.218	
	Mason	2	0	0	2	1	K	1.194	.671	−.523	
▲	Jackson	3	0	1	0	1	SF	1.390	1.243	−.147	▲
■	Jackson	0	2	0	1	2	1B	.102	.219	.117	■
	Jackson	0	2	0	0	3	FO	.102	0	−.102	
	Rivera	13	0	0	13	1	K	1.940	1.115	−.825	
	Andersen	1	0	0	1	1	GO	.813	.498	−.315	
	Greene	12	1	0	12	2	FO	.939	.403	−.536	
■	Greene	0	0	0	1	0	1B	.461	.813	.352	■
	Mason	0	0	0	0	1	FO	.461	.243	−.218	
	West	1	0	0	1	1	FO	.813	.498	−.315	
■	Andersen	0	1	0	1	1	1B	.243	.498	.255	■
■	Thigpen	0	0	0	2	0	2B	.461	1.194	.733	■
	Schilling	1	2	0	0	3	FO	.219	0	−.219	
	Schilling	1	1	0	1	2	K	.498	.219	−.279	
	Schilling	0	0	0	0	1	GO	.461	.243	−.218	
■ ▲	Schilling	0	0	1	0	0	HR	.461	1.461	1.000	▲ ■
▲	Mulholland	3	1	1	0	2	SF	.980	1.102	.122	▲
	Mulholland	0	1	0	0	2	GO	.243	.102	−.141	
	Mulholland	0	1	0	0	2	FO	.243	.102	−.141	
	Mason	0	0	0	0	1	FO	.461	.243	−.218	
	Williams	12	1	0	12	2	FO	.695	.318	−.377	

Sum of Change = −.902

TABLE 9-10   Change in Expected Runs for Joe Carter Plate Appearances in the "Parallel Universe," or "Twilight Zone" Version of the 1993 World Series (Squares Mark On-Base Events, Triangles Mark Plays That Scored Runs)

attention on the boldfaced rows. Everything in Table 9-10 is the same as in Table 9-9 except for these four rows, in which we have swapped the play results.

Suppose we swap the results of Carter's first two at-bats in Game 4. So, instead of singling with runners on first and second with 1 out in the first inning, Carter flies out without advancing the runners, and in the third inning he leads off with a single instead of flying out. In Table 9-10, we see that the run value of the fly out is now −.536 runs (instead of −.218 runs, as originally), while the single is now worth .352 runs (compared to .703 runs in reality). So, the swap produces a net change:

$$[-.536 - (-.218)] + (.352 - .703) = -.669 \text{ runs}$$

While we are making changes, let's swap that depressing (to us Phillies fans) HR in the ninth inning of Game 6 with Carter's fly out in his last at-bat in Game 5. Carter's second HR is now worth only 1 run (compared to 2.305 runs in reality) and the fly out in his final at bat in Game 6 is worth −.377 runs (compared to −.218 runs in its original spot).[9] So, the swap produces a net change as follows:

$$[-.377 - (-.218)] + (1 - 2.305) = -1.464 \text{ runs}$$

What we have done is preserve the count of individual batting events. In Table 9-10, Carter still has the same number of singles, doubles, triples, home runs, and outs as in Table 9-9. He still has the same batting average, on-base percentage, and slugging percentage. All we have done is change *the situation* in which they occurred. When we did this, the run values of the swapped events changed. If we total the run values in the Change column of Table 9-10, we see that Carter's performance in this alternate universe is now worth −.902 runs, a worse-than-average performance. We have substantially degraded Carter's run production merely by changing *when* the events occurred, rather than the *number* of each type of event. This example demonstrates that this measure successfully integrates a clutch effect (the *when* of batting results) into an evaluation of player performance.

A New Criterion for Performance

We've been able to integrate certain aspects of situational hitting into player evaluation. We have used the base and out aspects but (with the exception of situations in the bottom half of the ninth and extra innings) ignored the factors

---

[9] The expected runs scored with 2 outs and runners on first and second (trailing by 1 run in the bottom of the ninth inning) was determined in a manner similar to that for the same situation with 1 out. Using probabilities from Table 7-4, the value is calculated as:
$$(0 \times .791) + (1 \times .1) + [2 \times (.061 + .048)] = .318$$

which contribute to defining Late Inning Pressure—the inning and the score. To do this, we have to move from run production, which has been our major criterion for evaluating players, to a more general level.

The ultimate goal of any baseball team is not to score runs. Scoring runs is only the means to a higher goal, winning games. Teams can amass large run totals and still lose. Just look at the same 1993 World Series in which Carter played. In Game 4, the Phillies scored 14 runs and *still* lost to Toronto. And there is a classic example of the same phenomenon: Table 9-11 shows the scores for the seven games of the 1960 World Series, which the Yankees lost to the Pittsburgh Pirates.

The Yankees scored more runs in the first three games than Pittsburgh would score in the entire series. Overall, the Yankees scored more than twice as many runs as Pittsburgh. Despite this, the seventh game and the series were won by a Pirates homer that drove in only one run. There is no doubt that producing more runs increases your chances of winning. But a hit (even a single) at the right time can be more important to winning than a grand slam.

In 1970, a small book called *Player Win Averages: A Computer Guide to Winning Baseball Players,* by the brothers Eldon G. and Harlan D. Mills, developed a new metric for clutch play. The truly revelatory aspect of their system was that it focused on measuring not the events that lead to victory (the number of hits, walks, stolen bases, RBIs, etc.), but victory itself. But how can you measure a player's contribution to victory? According to the Mills brothers, you have to arrive at some sense of the degree to which the player contributes to the probability of winning a game.

Game	Pittsburgh	New York
1	6	4
2	3	16
3	0	10
4	3	2
5	5	2
6	0	12
7	10	9
Total	27	55

TABLE 9-11   1960 World Series Scores

They called this measure the *Player Win Average (PWA)*. It was defined as a ratio of Win Points to the sum of Win Points and Loss Points:

$$PWA = Win\ Points/(Win\ Points + Loss\ Points)$$

The points were based on how much the player added to or subtracted from the probability of his team winning. If a player *increased* the probability of his team winning by D, the player was awarded:[10]

$$Win\ Points = 2000 \times D$$

On the other hand, if a player *decreased* the probability of his team winning by D, the player was given:

$$Loss\ Points = 2000 \times D$$

Every play in a baseball game increases the probability of winning for one team or the other. Win and Loss Points were awarded on each play to an offensive player and a defensive player. A batter who got a hit was awarded Win Points based on how much the hit improved his team's probability of winning the game; the pitcher who gave up the hit was given an equal number of Loss Points. Naturally, the situation is reversed for an out, where a defensive player (typically the pitcher) is awarded the Win Points and the batter the Loss Points.

Consider this example, which the Mills brothers reckoned to be the biggest offensive play of the 1969 World Series, in which the Miracle Mets defeated the heavily favored Baltimore Orioles. Al Weis, the Mets' second baseman, came to the plate in the top of the ninth inning of Game 2 with the score tied. The Mets had runners on first and third with 2 outs, clearly a Late Inning Pressure situation. The probability of a Met victory was .510. Weis singled, placing runners at first and second as well as knocking in the go-ahead (and eventual game-winning) run. His hit raised the probability of a Met victory to .849, an increase of $D = .339$. Weis was therefore credited with the following Win Points:

$$Win\ Points = 2000 \times .339 = 678$$

Orioles pitcher Dave McNally was given an equal number of Loss Points.

In addition to analyzing the 1969 World Series, the Mills brothers used every play in the 1969 season to assign a PWA rating to every player. Each play in which a player was the primary offensive or defensive participant was analyzed. The Win Points and Loss Points from these plays were summed over the season, then substituted in the PWA formula. The highest PWA ratings for batters with enough plate appearances (502) to qualify for a batting championship were

---

[10] Since PWA is calculated as a ratio of points, it is not necessary to multiply the change in probability D by 2000 to get the same PWA value. Most likely the Mills brothers only used this conversion to make the Win and Loss Points easier to read.

achieved by Willie McCovey (.677) in the National League and Frank Robinson (.615) in the American League.[11] McCovey was selected by the sportswriters as National League MVP, but the American League MVP Award went to Harmon Killebrew, who had a .608 PWA.

When Win Points equal Loss Points, PWA equals .500, which is the rating for an average player. Clearly, McCovey, Robinson, and Killebrew were all way above average; their Win Points were all at least 50 percent greater than their Loss Points. But how low can PWA reasonably get? Bob Barton, a catcher with San Francisco, had a PWA of .255, lowest among players with 100 or more at-bats. Among full-time players, Hal Lanier, a shortstop also with the Giants, had a PWA of .348 in 495 at-bats, with almost twice as many Loss Points as Win Points. (The Giants appear to have had the best—McCovey—and worst PWA players in 1969.)

A unique capability of PWA is to measure defensive as well as offensive performance. In most plays, the pitcher is the defensive player who receives Win or Loss Points. The Mills brothers tabulated PWAs for pitchers in the exact same manner as for batters. The highest PWAs for starting pitchers were .612 for Larry Dierker of the Houston Astros (National League) and .585 for Denny McLain of the Detroit Tigers and Jim Palmer of the Baltimore Orioles (American League). The highest-rated relief pitchers were Tug McGraw (.651) of the New York Mets and Ken Tatum (.643) of the California Angels. McLain shared the 1969 Cy Young Award with Mike Cuellar (Baltimore Orioles, .569 PWA). Dierker, however, received no Cy Young votes, losing to Tom Seaver (New York Mets, .609 PWA).

Because both offensive and defensive players are measured according to the same metric, it is possible to use PWA to compare the value of pitchers and batters. McCovey is a clear standout when all players are considered, but the Mills' analysis indicates that a good case could have been made for Tatum over Killebrew as the most valuable player in the American League in 1969.

PWA also has the capability to include fielding in its evaluation of players. If a fielder made an error, the fielder was substituted for the pitcher in receiving Loss Points for the play. For example, in the bottom of the third inning of Game 1 in the 1969 World Series, Don Buford grounded a ball off Tom Seaver to second baseman Al Weis, who fumbled it, allowing Buford to reach first safely. Before

---

[11] Mike Epstein of the Washington Senators finished ahead of Robinson with a .641 PWA, but had only 500 plate appearances. Epstein was the Mills brothers selection as the "Most Winning" player in the American League in 1969.

the play, the Orioles led 1–0 and had no runners and one out in the bottom of the third inning. After the play, they had a runner on first, still with only one out. In the Mills' analysis, the play was split into two parts:

- *Win Points to Seaver*. Because Weis was charged with an error, the implication from the official scorer was that Buford would have been out if Weis had executed properly. So an intermediate result is created assuming that Weis had made the play. The result would have left the Orioles with two outs and no runners on base. Baltimore's probability of winning decreased from .658 to .644 for a change of D = .014. Seaver was awarded Win Points = 2000 × .014 = 28, and Buford was given 28 Loss Points.

- *Loss Points to Weis*. However, Weis's error reversed this intermediate result and placed a runner on first base with one out. This increased Baltimore's probability of winning from .644 to .679 for a change of D = .035. Buford was awarded Win Points = 2000 × .035 = 70, and Weis was given 70 Loss Points.

So the final result of the play was that Buford received 70 Win Points and 28 Loss Points, Seaver received 28 Win Points, and Weis received 70 Loss Points.

An unfortunate aspect of the Mills' book is that, apart from several examples, the authors do not provide detail about how they applied fielding in their analysis of the 1969 season. Although not explicitly stated, it is likely that fielding was included in the overall player ratings. So the PWA ratings cited are probably a reflection of fielding (and running) as well as batting performance.

The Mills brothers provided a play-by-play analysis of the 1969 World Series. An interesting result was that the true MVP of the series was not the sportswriters' choice, Donn Clendenon, but journeyman infielder Al Weis. Oddly, the authors based this selection not on PWA but on Net Points, the difference between Win Points and Loss Points. In many ways, this metric seems a more reasonable measure than PWA. Weis had 1277 Net Points, almost three times that of Clendenon (450 Net Points). In fact, three teammates, pitchers Koosman (783), Seaver (564), and Gentry (475), also had more Net Points than Clendenon. The second highest Net Point total was achieved by a member of the losing team, starting pitcher Mike Cuellar (998).

The Mills brothers used PWA to identify "Hidden Heroes," players who achieved above-average PWAs despite having low batting averages (less than .250). In the authors' view, these players were true clutch performers, those who rose above their normal ability when the game was on the line. Thirty years later, most of the players are not familiar, but one player does jump out. Hall-of-Famer Joe Morgan batted .236 in 1969, but he had a .521 PWA. Still, it is not clear that this PWA rating is a result of clutch performance. Morgan may have only had a .236 AVG, but he drew 110 walks, so his on-base percentage was quite respectable.

The Mills' goal in creating PWA was to evaluate clutch performance. Actually, PWA does *not* rate clutch performance; instead, it integrates or weights clutch performance with the frequency of different events. A good analogy is slugging percentage, which does not explicitly measure a player's power; instead, it integrates power (using the number of bases as weights) with batting average into the evaluation of a player's hitting performance. PWA does the same thing, except its weights are based on the game situation as well as the result of the play. Indeed, this is an even greater achievement than their original intention. Instead of rating one facet of a player's game, they established a structure for evaluating *all* aspects (including the player's clutch performance) into a single quantitative value.

Even so, as carefully thought out as their concept was, the Mills' work left some room for improvement. We have alluded to some areas already. In developing PWA, it is evident that the authors were intent on developing an average that could replace the batting average, which still reigned supreme and virtually unchallenged as the king of batting statistics in 1970. This is evident in the Mills' comparison of AVG with their statistic PWA for "Hidden Heroes."

Late in the book, it seems that the Mills brothers discovered that Net Points might be more useful in rating players according to contribution to winning. What are the advantages of Net Points?

- *Simplicity*. Net Points is easier to calculate than PWA. Just subtract Loss Points from Win Points, and you're done.

- *Intelligibility*. Dividing Net Points by 2000 gives the number of wins the player contributed above an average player. For example, 4000 Net Points is equivalent to two wins above average.

- *Consistency*. Consider Players A and B in Table 9-12. Both have the same Net Points, but different PWA ratings. Player A has a better PWA rating because the total number of points accumulated is less than that of Player B. There are two major ways Player A could have achieved this. One possibility is that he earned his points in less critical situations than Player B. In this way, each event would have less value—either positive or nega-

Player	Win Points	Loss Points	Net Points	PWA
A	2000	1000	1000	.667
B	3000	2000	1000	.600

TABLE 9-12   Example of Players with the Same Net Points and Different PWA Ratings

tive. It is not clear that this should entitle Player A to a higher rating than Player B. In fact, if PWA is truly supposed to evaluate clutch performance, it could be argued that Player B should be rated higher than Player A, since his results were achieved in more critical circumstances.

Another possibility is that Player A achieved his results in fewer plays (or games) than Player B. This presents some rationale for rating Player A higher than Player B. But then consider two more players, C and D, in Table 9-13. Player D has accumulated twice as many points as Player C, yet their PWA ratings are the same. So, dividing by total points does not give a consistent interpretation of player value in PWA.

In general, it is not clear what dividing by total points represents in evaluating player performance. On the other hand, Net Points provides a consistent interpretation of player contribution.

- *Sustained Contribution.* Apparently, the Mills' intent was to construct PWA as a ratio of accumulated achievement (Win Points) divided by accumulated opportunity (total points) as a parallel to batting average (which does the same thing in terms of hits and at-bats). However, in the Mills' system, when a player comes to bat, the possibility exists of getting Loss Points or Win Points. So in each play, the player can be rewarded or penalized. Net Points provides a measure of the accumulated net contribution to victory. For this reason, there is no need to resort to a ratio such as PWA. This can't be done with hits in batting average

Player	Win Points	Loss Points	Net Points	PWA
C	1000	1000	0	.500
D	2000	2000	0	.500

TABLE 9-13   Example of Players with the Same Net Points and the Same PWA Ratings

or bases in slugging percentage. Hits and bases can only be accumulated; we can't subtract hits or bases from a player.

The title page of *Player Win Averages* contains the phrase "1970 Edition." Apparently, the authors intended to publish an annual analysis of each baseball season using PWA. Unfortunately, there was no 1971 edition. By the late 1970s, the 1970 (and only) edition of *Player Win Averages* was out of print, and remains so to this day. Other publications and papers have alluded to the book and the PWA concept. A common criticism is provided by Palmer and Thorn in their book *The Hidden Game of Baseball*. "The major flaw in the Mills brothers' system is that the Player Win Average weights a few events very heavily, many others quite lightly . . ." (p. 176).[12] The impression is that PWA gives too much weight to a handful of critical events that drown out the effects of standard plays in evaluating baseball performance. Experience with a variant of PWA (to be described presently) in evaluating World Series performance over several years has provided evidence counter to this view.

Besides such criticism, another more practical reason lies at the heart of the lack of interest in *Player Win Averages* at the time. The Mills brothers provided no description of how to calculate the probability of victory at different stages of a baseball game. Using computer simulation, the brothers developed a table of probabilities that was not revealed to readers. Only the win probabilities of selected situations that arose in the 1969 World Series could be gleaned from the book. Without this table, it was impossible to calculate Win and Loss Points; without Win and Loss Points, PWA could not be calculated. So without the table of win probabilities, no one could use the technique. The need to capture play-by-play data is a large impediment to PWA's practicality, but the lack of win probabilities made its calculation impossible for anyone but the Mills brothers . . . until 1984.

The Calculation of Win Probabilities

Following the Phillies first and only World Championship, in 1980, John Flueck and one of us (Jay Bennett) wanted to see if Mike Schmidt was really the Most Valuable Player in the series. PWA seemed an ideal metric for determining this,

---

[12] Despite this perception, their succeeding volume (with Bob Carroll), *The Hidden Game of Football,* proposed Win Probability, which utilized basic principles similar to those of PWA. Win Probability was the probability of a football team winning the game based on field position, score, and time remaining. Win Probability points were credited to or subtracted from the offense and defense according to the change in Win Probability for different events. Sadly, *Player Win Averages* was not cited in the book.

but the effort was stymied because of the lack of the table of win probabilities. However, recent research in run production models led to Lindsey's data. We have already described some of this data, the distribution of runs scored after a given situation within an inning. Using data on the distribution of runs scored in each inning and assuming independence, Lindsey calculated the expected probability of winning given the score at the end of an inning, as shown in Table 9-14.[13]

Plotting these probabilities in Figure 9-2, we can see some quantitative support for the critical nature of Late Inning Pressure. Each line in Figure 9-2 represents a score differential for the home team over the visiting team. Looking at the lines in which the home team has the lead, each line curves upward as the game progresses. The smaller the lead, the more extreme the curvature. (The value of an extra run increases as the lead becomes smaller, and this value becomes even greater in late innings.) The difference between being ahead by 1 run or behind by 1 run after the first inning is about .2 in win probability, but about .7 in win probability after eight innings.

	Inning								
H–V	1	2	3	4	5	6	7	8	9
–6	.061	.054	.039	.029	.020	.012	.005	.0005	0
–5	.095	.087	.067	.053	.038	.025	.013	.004	0
–4	.142	.133	.109	.091	.070	.050	.029	.011	0
–3	.207	.196	.171	.150	.122	.093	.060	.025	0
–2	.290	.280	.257	.236	.207	.168	.122	.063	0
–1	.389	.383	.368	.353	.331	.295	.244	.153	0
0	**.500**	**.500**	**.500**	**.500**	**.500**	**.500**	**.500**	**.500**	.5
1	.611	.617	.632	.647	.669	.705	.756	.847	1
2	.710	.720	.743	.764	.793	.832	.878	.937	1
3	.793	.804	.829	.850	.878	.907	.940	.975	1
4	.858	.867	.891	.909	.930	.950	.971	.989	1
5	.905	.913	.933	.947	.962	.975	.987	.996	1
6	.939	.946	.961	.971	.980	.988	.995	.9995	1

TABLE 9-14   Probability of Home Team Victory Given the Score Difference (Home Team Minus Visiting Team) at the End of Each Inning

[13] Table 9-14 presents a slight revision of Table 7 in G. R. Lindsey, "The Progress of the Score During a Baseball Game," *American Statistical Association Journal*, September 1961, 703–728.

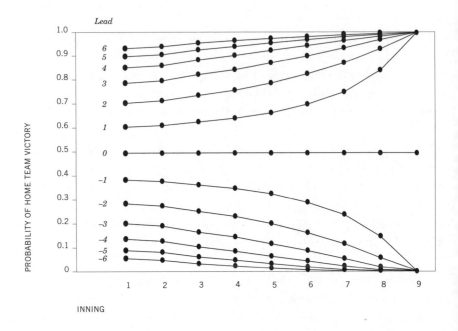

FIGURE 9-2   Probability of home-team victory given the home-team lead (home team minus visiting team) at the end of each inning.

So, Table 9-14 and Figure 9-2 capture the score and inning aspects of winning, while Table 7-4 captures the base and out aspects of scoring runs. Putting them together, we can derive a reasonable replication of the win probabilities used by the Mills brothers. The calculation to do this is somewhat like that for expected run production described earlier.

Suppose we wish to calculate the probability of winning when the home team trails by 1 run with runners on first and second and 1 out in the bottom of the ninth inning. This is the situation that Joe Carter faced as he approached the plate in the ninth inning of Game 6 in the 1993 World Series. Lindsey's data in Table 7-4 tells us the probability of scoring different numbers of runs in the remainder of the inning:

- *Toronto scores no more runs.* In this case (first and second base occupied, with 1 out), there is a .571 chance that Toronto will score no more runs in the remainder of the inning; if this happens, Toronto loses the game. The probability of a Toronto victory is 0 (as shown in Table 9-14).

- *Toronto scores 1 more run.* Table 7-4 also tells us that there is a
  .163 chance that Toronto will score exactly 1 more run in this
  inning. In this case, the game is tied and goes into extra innings.
  Toronto would then have an even chance (.5 probability) of
  winning (as shown in the ninth inning column of Table 9-14).

- *Toronto scores two more runs.* Table 7-4 tells us that there is a
  .119 chance that Toronto will score exactly 2 more runs in this
  inning. In this case, Toronto wins and its probability of victory is 1
  (as shown in the ninth-inning column of Table 9-14).

- *Toronto scores 3 or more runs.* Table 7-4 tells us that there is a
  .147 chance that Toronto will score 3 or more runs in this inning.
  Again, Toronto wins and its probability of victory is 1.

Table 9-15 summarizes the calculation. Each line of the table represents the possibility of scoring a specific number of runs in this inning given the situation (1 out and runners on first and second). The line then goes on to analyze the consequences of scoring those runs given the inning and the score. In the last column, we multiply the probability that the home team scores the runs times the probability that they win if they do. Summing the results in the last column gives the final result of the probability of winning given all that could possibly occur in the remainder of the inning. In this case, the probability of Toronto winning as Carter stepped to the plate was almost .35.

Let's look at a more general calculation, where the inning results are not win, lose, or tie the game, but where the Pr(Win Given Runs) values (other than tie) are different from 0 and 1. In *The Sporting News* #18 greatest moment, Sandy

SCORING IN INNING		RESULT OF SCORING		
Runs	Pr(Runs)	Home Lead	Pr(Win Given Runs)	Pr(Runs) × Pr(Win Given Runs)
0	.571	−1	0	0
1	.163	0	.5	.0815
2	.119	1	1	.1190
3 or more	.147	2 or more	1	.1470

Pr(Win) = .3475

TABLE 9-15   Calculation of the Probability of Home-Team Victory When the Team Trails by 1 Run in the Bottom of the Ninth Inning with 1 Out and Runners on First and Second Bases

Amoros of the Brooklyn Dodgers made his game-saving catch in the bottom of the sixth inning with his Dodgers ahead of the Yankees 2–0 in Game 7 of the 1955 World Series. Yogi Berra had come to the plate with runners on first and second, no outs. Table 9-16 summarizes the calculation of the Yankee probability of winning at the moment Berra came to bat. The Pr(Run) probabilities in the second column are taken from Table 7-4 for runners on first and second and no outs. The Pr(Win Given Runs) probabilities in the fourth column are taken from Table 9-14 for the sixth inning and the appropriate lead. The probability of a Yankee win was .3758, slightly better than the situation Toronto was in when Carter came to bat.[14] Comparing the situations, both teams had identical bases occupied. While New York trailed by more runs than Toronto, the Yankees had fewer outs in the inning and were not in their last at-bat, all of which more than counterbalanced their disadvantage in the score relative to the Blue Jays. All of these factors are accounted for in the calculation of the probability of winning.

We also see that there is more spread in the Pr(Win Given Runs) for Carter's situation (from 0 to 1) than for Berra's situation (from .168 to .705). This indicates greater instability in Carter's situation; that is, the possible change in victory from scoring 0 or 3 runs in the inning has a greater effect in Carter's situa-

SCORING IN INNING		RESULT OF SCORING		
Runs	Pr(Runs)	Home Lead	Pr(Win Given Runs)	Pr(Runs) × Pr(Win Given Runs)
0	.395	–2	.168	.0664
1	.220	–1	.295	.0649
2	.131	0	.500	.0655
3 or more	.254	1 or more	.705	.1791

Pr(Win) = .3758

TABLE 9-16   Calculation of the Probability of Home-Team Victory When the Team Trails by 2 Runs in the Bottom of the Sixth Inning with No Outs and Runners on First and Second Bases

[14] This calculation is actually a slight underestimate of the true value, which is closer to .397. In order to achieve this more accurate estimate, assumptions about the probability of scoring 3, 4, 5, 6, etc., runs must be made to replace the single "3 or more" value given by Lindsey. The calculation then follows the same pattern outlined here. The calculation of the Carter situation is exact because the probability of a Toronto win is 1 as long as 2 or more runs score.

tion than in Berra's, and this indicates that the degree of clutch is more intense in Carter's than in Berra's.

Applying this system to Lindsey's data, Bennett and Flueck were able to calculate their own table of win probabilities. Figure 9-3 is an application of these win probabilities to Game 6 of the 1993 World Series. The line tracks the probability of Toronto victory as it changed after each play. The effects of plays in which Joe Carter batted are emphasized with heavy lines. The end of each inning is indicated on the x-axis, giving the score at the end of the inning (Toronto– Philadelphia). A diamond symbol on the line indicates the result of the last play of the visitors' (Phillies') half of the inning.

We see that Toronto's probability of winning rose from the start as they prevented the Phillies from scoring in the first inning. This trend continued as the Blue Jays scored three times in the first inning. Carter's sacrifice fly was part of this effort; it provided a small boost in the probability by knocking in the second run of the inning. There was no scoring in the next two innings. Toronto's probability of winning rose slightly as they preserved their lead. The fourth inning saw a small dip in Toronto's probability of victory as the Phillies scored a run. By

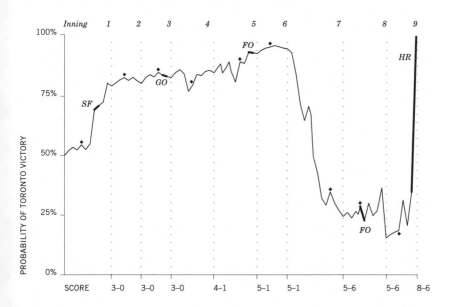

FIGURE 9-3   Win probabilities for Toronto after each play in Game 6 of the 1993 World Series. Effects of plate appearances by Joe Carter are emphasized with heavy lines. The scores at the end of each inning are marked on the x-axis. Diamonds identify result of last play in visitors' (Phillies') at-bat.

matching Philadelphia with a run of their own, Toronto restored their three-run lead and inched their probability of winning to an even higher level. The Phillies had an opportunity to score some runs in the fifth inning (bases loaded and 2 outs), but failed when Dave Hollins grounded into the third out. The low point in the inning for Toronto occurred just after John Kruk walked to load the bases. Note how win probability is able to account for establishing a scoring threat even if it eventually fails. Hollins's big out is shown by the rise (ending with the diamond symbol) in Toronto's victory probability from this low point in the fifth inning. Toronto responded to this scare with another run, to inch ahead their lead and chances of winning.

The sixth inning was uneventful, but in the seventh the Phillies turned the tables on Toronto, scoring five runs to take the lead. The plot of Toronto's probability of winning looks like the Dow Jones Industrial Average in free-fall. Toronto did not score in its half of the seventh, pushing its probability of winning even further down. When the Phillies did not score in the top of the eighth inning, Toronto had a chance to tie or get the lead. Joe Carter had Toronto's first crack at the Phillies, but flied out in a critical at-bat that decreased Toronto's chances even further. After Carter, the plot rises (and the plot thickens) in the eighth inning as a result of a threat with bases loaded and two outs. Like Hollins on the opposing team, Pat Borders made the third out, producing a net gain in the probability of the Phillies winning in the inning since their one-run lead was preserved. In the ninth inning, the Phillies went down quickly. Toronto came out storming in the ninth, quickly putting runners on base. The triumphant (or cataclysmic, depending on your view) impact of Carter's home run is evident in the steep rise in the final markings of the plot.

Player Game Percentage (PGP)

In addition to the calculation of win probability, Bennett and Flueck made other modifications. They adopted the Net Points viewpoint instead of the PWA ratio concept. Instead of using Win Points and Loss Points, their measure used the change in probability directly expressed as a percentage. Half of the change was attributed to the offensive player's performance, and the other half to the defensive player's performance. This simplification allowed the direct computation of wins above average without the need for dividing by 2000 (as in the Mills system).

Consider Joe Carter's home run in the 1993 World Series finale. As described in Table 9-15, the probability of a Toronto victory was .3475 when Carter came to bat. His home run won the game, so the probability of a Toronto win was exactly 1 after his at-bat. The change in Toronto's win probability was $1 - .3475 = .6525$, or 65.25 percent. Half of this change (32.63 percent) was awarded to Carter, and the negative half of the change (–32.63 percent) was given to Mitch Williams, the hapless Phillies reliever who delivered the ill-fated pitch.

For the most part, Bennett and Flueck adopted the procedures used by the Mills brothers in identifying the major offensive and defensive contributors in each play, following the examples provided by the Mills' analysis of the 1969 World Series. There were departures, however. One of these was the method used to evaluate errors. Earlier in this chapter, we looked at the Mills brothers' analysis of Weis's error in Game 1 of the 1969 World Series. Basically, the analysis broke the play up into two parts: a ground-out giving Win Points to the pitcher Seaver, and the error giving Loss Points to Weis. However, the batter ended up getting 28 Loss Points on the ground ball to Seaver and 70 Win Points on Weis's error, for a net gain of 42 points, more than the 28 Win Points given to Seaver. It did not seem right for the batter to get any positive recognition for this play, much less greater recognition than the pitcher.

The solution adopted by Bennett and Flueck was not to give any positive credit to the batter in this play and award the entire negative change from the error to the fielder. So, using the example above, the first part of the play is the same (except for the change from the Points framework); the change was $D = .014$, or 1.4 percent, so Seaver gets $D/2 = 0.7$ percent, and Buford gets –0.7 percent. However, Weis's error produced a negative change (from his team's perspective) of $D = .035$, so he is debited the entire change, –3.5 percent. In this way, only the pitcher receives any positive recognition from the play.

PGP (and PWA) have very powerful capabilities to quantify defensive contributions to winning. However, while the mechanics of the probability calculations are objective, identifying the players and whether their defensive contributions were outstanding enough for special recognition remain subjective judgments. This is less of a problem for errors, since MLB has assigned official scorers the task of identifying misplays in the field. But how do we identify great plays by fielders? And once recognized, how do we reward the fielder? Willie Mays' renowned catch in the 1954 World Series, for example, was unquestionably a great fielding play. One possibility is just to give the defensive credit for the out to the fielder instead of the pitcher, as might be done in the case of Mays' catch, but is this enough credit for an extraordinary play? Probably not. So, the other possibility is to analyze it as an error in reverse. Split the play into two parts, the first being a hit off the pitcher and the second being the out credited to the

				BEFORE PLAY				AFTER PLAY				
D	Play	Game	Inning	Lead	Bases	Outs	Pr(Win)%	Lead	Bases	Outs	Pr(Win)%	
■ 3.80	1B	1	2H(L)	−2	0	0	32.91	−2	1	0	36.71	■
▲ 2.07	SF	1	3H	−1	3	1	49.16	0	0	2	51.23	▲
−3.22	GO	1	6H(L)	0	0	0	57.16	0	0	1	53.94	
−.22	K	1	7H	4	2	1	98.41	4	2	2	98.19	
−2.15	FO	2	1H	0	1	2	52.15	0	0	3	50.00	
■▲ 9.68	HR	2	4H	−5	1	0	10.10	−3	0	0	19.78	▲■
−2.49	FO	2	6H(L)	−3	0	0	14.33	−3	0	1	11.84	
−6.67	K	2	8H	−3	2	0	15.43	−3	2	1	8.76	
▲ −.31	SF	3	1V	2	3	0	76.89	3	0	1	76.58	▲
■ .44	1B	3	3V	4	0	2	85.92	4	1	2	86.36	■
−.37	FO	3	5V	4	0	2	89.81	4	0	3	89.44	
−.62	K	3	7V	5	13	0	98.77	5	13	1	98.15	
−.04	GO	3	9V	6	1	0	99.75	6	1	1	99.71	
■ 6.64	1B	4	1V	0	12	1	54.67	0	123	1	61.31	■
−2.40	FO	4	3V(L)	−3	0	0	19.60	−3	0	1	17.20	
−2.23	FO	4	4V(L)	1	0	0	63.20	1	0	1	60.97	
−2.34	FO	4	6V	−4	1	0	9.69	−4	1	1	7.35	
■ .61	1B	4	8V	−5	0	1	.89	−5	1	1	1.50	■
■ 4.78	2B	4	9V(L)	1	0	0	84.70	1	2	0	89.48	■
−2.20	FO	5	1V	0	1	2	47.69	0	0	3	45.49	
−3.23	K	5	4V	−2	1	1	26.29	−2	1	2	23.06	
−3.38	GO	5	7V(L)	−2	0	0	16.80	−2	0	1	13.42	
−3.01	FO	5	9V(L)	−2	0	0	6.30	−2	0	1	3.29	
▲ 1.45	SF	6	1H	1	3	1	69.38	2	0	2	70.83	▲
−.69	GO	6	3H	3	0	1	83.94	3	0	2	83.25	
−.32	FO	6	5H	4	0	1	93.58	4	0	2	93.26	
−6.03	FO	6	8H(L)	−1	0	0	28.67	−1	0	1	22.64	
■▲ 65.25	HR	6	9H	−1	12	1	34.75	2	0	3	100.00	▲■

TABLE 9-17   Player Game Percentage Evaluation of Joe Carter's Performance in the 1993 World Series in Chronological Order (Squares Mark On-Base Events, Triangles Mark Plays That Scored Runs)

fielder. PGP has used both methods at times for evaluating fielding plays, but there is much room for improvement in this aspect of its use.

While Bennett and Flueck's measure was derived from the same concepts as PWA, there were enough differences (especially the use of their own table of win probabilities) that they gave their measure its own name, *Player Game Percentage (PGP)*.

Let's see how well PGP rated Carter's 1993 World Series performance. Table 9-17 is the same as Table 9-8 except that it also presents the probability of a Toronto victory before and after each play (as a percentage). The first column (D) gives the change in these probabilities, subtracting the Before probability from the After probability. Carter's net contribution over the course of the series can be found by summing the values in the D column and dividing by 2 (since the change is split in half between offense and defense). This calculation produces the following net contribution:

$$52.80/2 = 26.40$$

This is roughly equivalent to a quarter of a win over the course of the six games.[15] Carter's PGP rating is then found by dividing his net contribution by the number of games:

$$PGP = \frac{Net\ contribution}{Number\ of\ games} = \frac{26.4}{6} = 4.4$$

This means that Carter's play in the 1993 World Series was good enough to raise a team's winning percentage by .044. So, a play of this caliber over the course of a season could raise a .500 team to a .544 team.

We can summarize Carter's play game-by-game by just adding up his contributions in each game individually. Figure 9-4 displays these values in the bars. The line shows Carter's PGP for the series as it progresses; each value is the sum of the Game PGPs divided by the number of games to that point in the series. We see that Carter's PGP fluttered about the average (0) level until the big boost in Game 6.

Looking at Table 9-17, we see that Carter got off to a good start in Game 1, with major contributions early in the game. His sacrifice fly tied the game, which Toronto would eventually win. Game 2 saw the best (almost) and worst of Carter's play. His home run chipped into the early Phillies lead, and was Carter's

---

[15] Before his final home run, Carter actually had a *negative* net contribution ($-12.45/2 = -6.225$). A negative value does not mean that he prevented a victory but rather that his performance was below average up to that point.

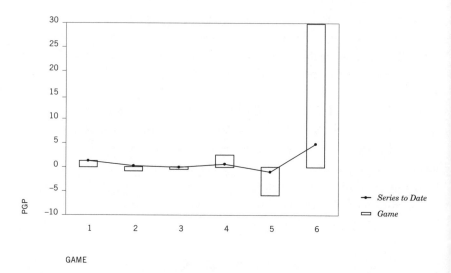

FIGURE 9-4   Game history of Joe Carter's Player Game Percentage throughout the 1993 World Series.

second biggest play of the series. On the other hand, he struck out late in the game, putting the brakes on a possible big inning to get Toronto back in the game; this was his worst at-bat in the series. Despite the HR, his contribution was negative in this game. His contributions in Game 3 were relatively minor (positively and negatively) in a Toronto victory that was assured early in the game.

Game 4 was a wild affair that Toronto won 15–14. Carter played a positive role in Toronto's victory. His first single set the stage for a big Toronto first inning. His second single started the rally that eventually brought Toronto from way behind into the lead, but it gave Carter very little credit (+.3). When the single was hit, the likelihood of a Toronto comeback was very remote, a probability of winning less than .01 before the single and less than .02 after it. Toronto had to score 5 runs just to tie and thus reach a 50-percent chance of winning. The view that Carter's single was a big hit is hindsight; PGP (and PWA) evaluate a play at the moment of its resolution, not after the fact. The credit for the Toronto rally went to the hits made by Rickey Henderson and especially Devon White, who delivered when the game was more in doubt, later that inning. Ironically, Carter got more credit for his double in the ninth inning, setting up a potential insurance run (which did not score). PGP and PWA operate on probabilities, but sometimes the remote possibilities *do* occur.

Game 5 was Carter's worst game, as it was for the rest of Toronto's players. Phillie starting pitcher Curt Schilling took charge and pitched a shutout. Game 6 started off well for Carter as his sacrifice fly helped Toronto to an early lead. But unlike Game 4, when his team, trailing by a lone run, needed a runner, he flied out to lead off the eighth inning; this was his second greatest negative at-bat of the series, although his final at-bat more than made up for it.

The analysis of Carter's at-bats within the context of the game using probability of victory lets us see the ebb and flow of player production. It quantifies our intuitive feel for the great dramatic plays such as Carter's home run to win the final game and the series. Unlike most statistical summaries of player performance, a PGP/PWA analysis tells a revealing story.

Suppose we sort Table 9-17 not chronologically but according to play type, as shown in Table 9-18. What does this tell us about the value of different plays? Of course, we must keep in mind that this is a very small (and unrepresentative) sample of situations in which these plays could occur. Still, we see that the evaluation of these plays follows our intuition of their values. The average value of Carter's home runs (average $D = 37.5$) is higher than that of any other play type. The average value of a single (average $D = 2.9$) is less than that of the double ($D = 4.8$) and greater than that of the sacrifice fly (average $D = 1.1$). All outs (except SFs) have negative values.

However, taking a closer look beyond the averages, we notice exceptions to these generalities. One single which energized a big opening in Game 4 had greater value than Carter's lone double. Two sacrifice flies had more value than some singles; both SFs occurred with one out and the score close. On the other hand, one SF which occurred with no outs had a *negative* value; apparently, an SF with no outs reduces the possibility of a big inning, while an SF with one out in a close game guarantees the scoring of a run when less opportunity exists. Also note that Carter's leadoff outs tended to be more damaging than other outs. In fact, the average value of Carter's leadoff outs was −3.3, and −1.7 for all other outs, excluding SFs. This provides some quantitative support for the leadoff spot being a critical element of each inning.[16]

---

[16] In general, given the same score and inning, leadoff outs are more costly than outs in bases-empty situations with 1 or 2 outs already achieved, but less costly than outs in most situations with runners on base.

	D	Play	Game	Inning	BEFORE PLAY				AFTER PLAY				
					Lead	Bases	Outs	Pr(Win)%	Lead	Bases	Outs	Pr(Win)%	
■ ▲	65.25	HR	6	9H	−1	12	1	34.75	2	0	3	100	▲ ■
■ ▲	9.68	HR	2	4H	−5	1	0	10.10	−3	0	0	19.78	▲ ■
■	4.78	2B	4	9V(L)	1	0	0	84.70	1	2	0	89.48	■
■	6.64	1B	4	1V	0	12	1	54.67	0	123	1	61.31	■
■	3.80	1B	1	2H(L)	−2	0	0	32.91	−2	1	0	36.71	■
■	.61	1B	4	8V	−5	0	1	.89	−5	1	1	1.50	■
■	.44	1B	3	3V	4	0	2	85.92	4	1	2	86.36	■
▲	2.07	SF	1	3H	−1	3	1	49.16	0	0	2	51.23	▲
▲	1.45	SF	6	1H	1	3	1	69.38	2	0	2	70.83	▲
▲	−.31	SF	3	1V	2	3	0	76.89	3	0	1	76.58	▲
	−.04	GO	3	9V	6	1	0	99.75	6	1	1	99.71	
	−.22	K	1	7H	4	2	1	98.41	4	2	2	98.19	
	−.32	FO	6	5H	4	0	1	93.58	4	0	2	93.26	
	−.37	FO	3	5V	4	0	2	89.81	4	0	3	89.44	
	−.62	K	3	7V	5	13	0	98.77	5	13	1	98.15	
	−.69	GO	6	3H	3	0	1	83.94	3	0	2	83.25	
	−2.15	FO	2	1H	0	1	2	52.15	0	0	3	50	
	−2.20	FO	5	1V	0	1	2	47.69	0	0	3	45.49	
	−2.23	FO	4	4V(L)	1	0	0	63.20	1	0	1	60.97	
	−2.34	FO	4	6V	−4	1	0	9.69	−4	1	1	7.35	
	−2.49	FO	2	6H(L)	−3	0	0	14.33	−3	0	1	11.84	
	−2.40	FO	4	3V(L)	−3	0	0	19.60	−3	0	1	17.20	
	−3.01	FO	5	9V(L)	−2	0	0	6.30	−2	0	1	3.29	
	−3.22	GO	1	6H(L)	0	0	0	57.16	0	0	1	53.94	
	−3.23	K	5	4V	−2	1	1	26.29	−2	1	2	23.06	
	−3.38	GO	5	7V(L)	−2	0	0	16.80	−2	0	1	13.42	
	−6.03	FO	6	8H(L)	−1	0	0	28.67	−1	0	1	22.64	
	−6.67	K	2	8H	−3	2	0	15.43	−3	2	1	8.76	

TABLE 9-18   Player Game Percentage Evaluation of Joe Carter's Performance (Sorted by Play Type) in the 1993 World Series (Squares Mark On-Base Events, Triangles Mark Plays That Scored Runs)

World Series Most Valuable Players

Continuing the tradition established in *Player Win Averages*, the first application of PGP was to evaluate players in the 1980 World Series. Since then, PGP has been used to analyze each World Series since 1987. Just as the Mills brothers identified the erroneous selection of Donn Clendenon as the 1969 World Series MVP over Al Weis, it has been interesting to see how often the sportswriters' MVP selection matched the PGP selection.

Despite his spectacular home run, Carter did not have the highest PGP rating in the 1993 World Series. That honor belonged to Paul Molitor (6.8 PGP), who was the sportswriters' pick as the Series Most Valuable Player. The two honors do not always coincide. Table 9-19 presents the official MVP selected by the sportswriters and the unofficial MVP selected by PGP. Indeed, at times, PGP has selected a player from the *losing* team! These players (Aikens, Pena, Jones, and Gwynn) deserved a better fate. Some MVPs actually had subpar performances, as indicated by their negative PGP ratings.

With two exceptions, the PGP ratings presented for pitchers here did not include their appearances at bat. PGP rates players with respect to average per-

TEAMS			SPORTSWRITERS' MVP		HIGHEST PGP	
Year	Winner	Loser	Player	PGP	Player	PGP
1980	Philadelphia	Kansas City	Mike Schmidt, 3b	4.6	**Willie Aikens, 1b**	8.0
1987	Minnesota	St. Louis	Frank Viola, p	.8	**Tony Pena, c**	2.8
1988	Los Angeles	Oakland	Orel Hershiser, p	5.4	Kirk Gibson, ph	8.7
1989	Oakland	San Francisco	Dave Stewart, p	−.8	Mike Moore, p	6.2
1990	Cincinnati	Oakland	Jose Rijo, p	8.5	Jose Rijo, p	8.5
1991	Minnesota	Atlanta	Jack Morris, p	7.6	Jack Morris, p	7.6
1992	Toronto	Atlanta	Pat Borders, c	−.6	Ed Sprague, ph–1b	5.5
1993	Toronto	Philadelphia	Paul Molitor, dh–1b	6.8	Paul Molitor, dh–1b	6.8
1995	Atlanta	Cleveland	Tom Glavine, p	5.1	Tom Glavine, p	5.1
1996	Yankees	Atlanta	John Wetteland, p	3.6	**Chipper Jones, 3b**	3.7
1997	Florida	Cleveland	Livan Hernandez, p	−.9	Gary Sheffield, of	3.2
1998	Yankees	San Diego	Scott Brosius, 3b	7.0	**Tony Gwynn, of**	8.1
1999	Yankees	Atlanta	Mariano Rivera, p	3.6	Chuck Knoblauch, 2b	4.8
2000	Yankees	Mets	Derek Jeter, ss	3.8	Mariano Rivera, p	5.7

TABLE 9-19   Most Valuable Players in Recent World Series (Players in Boldface Were on the Losing Team)

formance. Pitchers are at a severe disadvantage compared to an average hitter, and their ratings generally suffer accordingly. Since pitchers are not expected to hit, their hitting skills are considered a bonus rather than an expectation. This is especially relevant in recent World Series play, in which some games have pitchers batting and others do not because of the designated hitter rule. The exceptions are Orel Hershiser and Mike Moore. Hershiser's 3 for 3 in the 1988 World Series raised his 4.9 PGP rating from pitching alone to an overall PGP rating of 5.4. Mike Moore's two-out double early in Game 4 of the 1989 World Series, which knocked in 2 runs, was more valuable than his pitching performance (which was excellent in its own right). His batting performance raised his overall PGP rating to 6.2 from his pitching PGP rating of 4.2. It would be ignoble to ignore these valuable (unexpected) contributions when highlighting great performances in the World Series.

The sportswriters selected a pitcher as MVP in 9 out of 14 World Series examined, while PGP selected a pitcher only 5 times. From the PGP viewpoint, the media are not properly appreciative of the relatively rare but powerful contributions made by offense to team victory. Nowhere is this more evident than in the evaluation of the 1988 World Series. Kirk Gibson came to bat only once in the series, but in that one at-bat he turned defeat into victory, raising the probability of victory from about .13 to 1. Orel Hershiser pitched very well in two starts, as indicated by his high PGP rating. However, in both games, the Dodger offense produced early leads (5–0 after three innings in Game 2, and 2–0 before Hershiser threw his first pitch in Game 5). Hershiser's contribution was mainly to preserve a lead already given to him. Pitching cannot win a game alone, as Gibson's home run did. From the perspective of winning games, Gibson's lone but seismic contribution (#6 among the game's greatest moments in the opinion of *The Sporting News*) was greater than Hershiser's accomplished (but not critical) performance.

Critics of the PWA/PGP system might say "Aha! We told you that too much weight is given to the big hit!" If that is so, how do they explain that Tony Gwynn's performance in the 1998 World Series (on a team that lost in four straight!) is rated higher by PGP than the performance of Scott Brosius. After all, Brosius had the biggest hit of the series, a home run in Game 3 that brought the Yankees from behind into the lead late in the game. And it was his second home run of the night. (His first, only an inning earlier, started the Yankee comeback.) Figure 9-5 shows PGP ratings for Brosius and Gwynn in each game of the 1998 World Series. We see that Game 3 was definitely the highlight of Brosius' series; he also had a great performance in Game 2, but Games 1 and 4 were

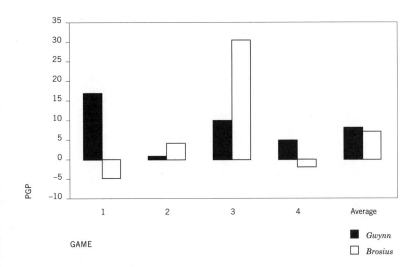

FIGURE 9-5   PGP ratings of Tony Gwynn and Scott Brosius in the 1998 World Series.

below average. Gwynn, on the other hand, had no dramatic game-winning moment (how could he, when the Padres were swept in the series?), but he had consistently good performances; his PGP rating was positive in *every game of the series*. Apparently, PGP is capable of rewarding consistently good play that contributes to a team's *chances* of winning—even if the victory afforded by the opportunity presented is not realized. This view is further buttressed by Molitor being rated higher than Carter in the 1993 World Series despite Carter's game-winning HR.

Comparing PGP ratings in Table 9-19, one might get the impression that since Kirk Gibson had the highest PGP rating, his performance must be rated the best of all players in recent World Series. However, when comparing performances in different series, it is better to use Net Contribution.

$$Net\ contribution = PGP \times number\ of\ games$$

Remember that with each play a player's Net Contribution can go down as well as up, so playing in more games is not necessarily an advantage. Listing the best players from each series according to Net Contribution moves Jack Morris's pitching performance in the 1991 World Series (climaxed by his heroic 11-inning shutout victory in Game 7) to the top of the heap in Table 9-20. Willie Aikens's name would undoubtedly be better remembered if his Kansas City Royals had prevailed over the Phillies in 1980. His high Net Contribution indicates that his

Year	Player	PGP	Games	Net Contribution
1991	Jack Morris, p	7.6	7	53.2
1980	Willie Aikens, 1b	8.0	6	48.0
1988	Kirk Gibson, ph	8.7	5	43.5
1993	Paul Molitor, dh-1b	6.8	6	40.8
1990	Jose Rijo, p	8.5	4	34.0
1992	Ed Sprague, ph-1b	5.5	6	33.0
1998	Tony Gwynn, of	8.1	4	32.4
1995	Tom Glavine, p	5.1	6	30.6
2000	Mariano Rivera, p	5.7	5	28.5
1989	Mike Moore, p	6.2	4	24.8
1997	Gary Sheffield, of	3.2	7	22.4
1996	Chipper Jones, 3b	3.7	6	22.2
1987	Tony Pena, c	2.8	7	19.6
1999	Chuck Knoblauch, 2b	4.8	4	19.2

TABLE 9-20   Net Contributions of Highest PGP-Rated Players in Recent World Series

22 total bases in 20 at-bats (for an amazing 1.100 slugging percentage!) made contributions to victory that the rest of the team (primarily the pitching staff) could not preserve.

A common question that comes up in reference to World Series play is whether PGP can be adapted to evaluate players with respect to the probability of winning the series as opposed to winning individual games. This is certainly possible, although we refrain from using the system in this way. The basic unit of play in baseball is the game. Each player is focused on that objective. Evaluating players on that basis provides a uniformity and consistency of the application throughout the season or at any level in the playoffs.

It should be made clear that PGP and PWA as defined here are not measures of ability but of observed performance. In order to draw inferences about ability, they must be modeled using principles similar to those outlined in other chapters of this book.

Looking to the Future

This investigation started with ways to measure clutch play. But we may have accomplished much more. Instead of isolating clutch play and examining player capabilities with respect to that one facet alone, we have found a system that integrates clutch play into an overall evaluation of player performance: One which can be used to compare starting pitchers with relief pitchers, pitchers with hitters. One which recognizes the importance of a good bench in winning games. One which is able to evaluate stolen bases within the context of a game.

With the increasing availability of play-by-play records, evaluating players on the basis of probability of winning using metrics like PWA and PGP may be the wave of the future.

PREDICTION

CHAPTER 10

Part of what makes baseball an exciting sport, at least to us, is its unpredictability. During the preseason, experienced and knowledgeable sportswriters make a number of predictions regarding the best and worst teams, the best and worst hitters and pitchers. But it is pretty common for the season to play out in a way very different from what is predicted. Teams often perform much worse or better than expected, and players many times achieve much more (or less) than expected. Many fans can remember teams like the Whiz Kids (the 1950 Philadelphia Phillies) and the Miracle Mets (the 1969 New York team), who had amazingly successful seasons despite virtually unanimous preseason predictions of mediocrity. Likewise, it is easy to think of young players who surprised everyone with great seasons. (Remember Mark Fidrich of the Tigers and, more recently, Sean Casey of the Reds?) So it is clear that there is a lot of uncertainty in any preseason prediction. In this chapter, we will describe some statistical methods that can be used in prediction and make some comments about the accuracy of these predictions. We will first talk about prediction of individual game outcomes, then discuss prediction of home-run numbers for two recently famous and fantastically successful sluggers, Sammy Sosa and Mark McGwire.

Predicting Game Results

Let's first focus on the basic problem of predicting game results. Suppose team A, say Anaheim, plays team B, say Boston, in a single game at team A's home park. Can we, with any degree of certainty, answer the question, "Who will win?"

Guessing

The simplest way of predicting the outcome of a baseball game (or anything else) is to make a blind guess. Suppose we knew nothing about baseball and we thought that the teams all had roughly the same ability. Also, since we're ignorant about baseball, we are not aware of any advantage (to either team) of playing in team A's ball park. So we make our prediction by flipping a coin. If the coin lands heads, we'll predict A to win the game; if the coin is tails, we'll pick team B.

Suppose that we use this "guess method" in predicting many baseball games. How good are these predictions? Well, since the result of the coin flip has nothing at all to do with the outcome of the baseball game, we can expect to be right about half of the time. In other words, our success rate using the guess method would be 50 percent.

Picking the Home Team

We can certainly improve on mere guessing, since we do know something about baseball. For example, we know that there is some advantage for a team playing in their home ball park. So it would be reasonable to always predict that the home team would win. How good is this prediction? In 1999, 52.1 percent of all games were won by the home team. So our success rate if we predict the home team is 52.1 percent.

In other words, our predictions are slightly better if we pick the home team than if we simply guess, but the improvement is pretty small—only 2.1 percent. The home-field advantage in baseball is much smaller than the home-court effect in basketball or the home-field advantage in football. (In recent professional football games, the home team wins approximately 58 percent of the time, and in professional basketball the home team wins 66 percent.)

A "Team Strengths" Prediction Model

We can do better than merely choosing the home team. We know that teams have different abilities, and we should be able to use this knowledge to make a more accurate prediction. We will describe one simple statistical method, similar to the least-squares method of fitting a line to a scatterplot (as discussed in Chapter 5), and use it to make a prediction based on the relative strengths of the teams.

First, we will code the game results, with team A as the home team and team B the visitor. If the home team (A) wins, we will record the result as +1; if the visitor (B) wins, we'll record a −1. So all of the game results for the season are recorded as a sequence of +1s and −1s.

Next, we need to construct a prediction formula that takes into account the teams' strengths and the home-field advantage. We'll denote the strength of team A as S_A and the strength of team B as S_B, and let H denote the home-field advantage. A simple prediction formula says:

$$Prediction = S_A - S_B + H$$

If we knew the strengths of the two teams and the home-field effect, then we can compute Prediction. For example, if Anaheim had a strength of .3 and Boston had a strength of .4 and the home field advantage is .2, then our prediction would be:

$$Prediction = .3 - .4 + .2 = .1$$

We can then use the value of Prediction to make our game prediction as follows:

- If Prediction is positive, then Team A will win.
- If Prediction is negative, then Team B will win.

In our example, since Prediction is .1 (a positive number), we would predict Anaheim to win the ball game.

The only problem with this method is that we don't know either the team strengths or the home-field advantage. To use this prediction method, we have to estimate these numbers, using data from the season. We use a popular estimation method called least-squares to do this.

Suppose we collect the results for a large number of games. For each game we record the game outcome (+1 if the home team wins, −1 if the visiting team wins) and the names of the two teams that played. There are 30 teams—we don't know the strengths of the teams which we represent by $S_1, \ldots, S_{30}$, and we don't know the value of the home-field advantage H. Using the least-squares method, we find values of the team strengths and the home-field advantage which makes the sum of $[Outcome - (S_{home} - S_{away} + H)]^2$ as small as possible, where S_{home} is the strength of the home team and S_{away} is the strength of the visitor.

Predicting 1999 Game Results

Let's illustrate this method for predicting the results of the games in the 1999 regular season. At the beginning of the season, we can't make any predictions, since we don't know the team strengths. We could use the results from the 1998 season to learn about the team abilities, but there is a lot of movement of players between teams in the off-season, and it is not clear how relevant the 1998 game results will be in predicting 1999 results.

Team	Strength				
Anaheim	0	Detroit	–0.07	Oakland	0.13
Arizona	0.41	Florida	–0.11	Philadelphia	0.19
Atlanta	0.44	Houston	0.42	Pittsburgh	0.26
Baltimore	–0.17	Kansas City	0	San Diego	0.03
Boston	0.32	Los Angeles	0.29	San Francisco	0.29
Chi Cubs	0.36	Milwaukee	0.13	Seattle	0.08
Chi White Sox	0.04	Minnesota	–0.16	St Louis	0.25
Cincinnati	0.30	Montreal	0	Tampa Bay	–0.02
Cleveland	0.40	NY Mets	0.24	Texas	0.23
Colorado	0.16	NY Yankees	0.24	Toronto	–0.02

Home Field Advantage	0.01

TABLE 10-1   Team Strengths and Home-Field Advantage Estimated Using Game Results of First Two Months of the 1999 Season

So we will initially use the game results for the first two months of the season in 1999 to estimate the team strengths. From these game results, we use the least-squares method to obtain values of the team strengths and the home-field advantage. These least-squares estimates are displayed in Table 10-1.

Now we can use the least-squares method to predict the results of the games on June 1. Table 10-2 gives the predictions, the game results, and whether we were right or wrong in our prediction for this particular day.

Let's illustrate this prediction process for one game, the Atlanta-Colorado game played on June 1, 1999. We see from Table 10-1 that Atlanta has a strength of .44 and Colorado has a strength of .16, and the home effect is estimated to be .01. So for this game, the value of Prediction is calculated as follows:

$$\text{Prediction} = .44 - .16 + .01 = .29$$

This means that we predict Atlanta will win. From Table 10-2, we see that the home team (Atlanta) won this game, so we were right on this prediction.

For the next day (June 2), we repeat the process. Using the 1999 data for all of the games prior to June 2, we estimate the team strengths and the home advantage. We use the formula to predict the June 2 games and keep track of our correct and incorrect predictions. The next day we update our estimates of the team strengths and H using the new set of results through June 2, then using these values in the new formula to predict the June 3 results, and so on. How did we do? In predicting 1616 games during the 1999 season, the formula gave correct predictions for 924 games. In other words:

Home Team	Visiting Team	Prediction	Who Won Game?	Right or Wrong?
Anaheim	Minnesota	0.17	Home team	Right
Atlanta	Colorado	0.29	Home team	Right
Boston	Detroit	0.40	Home team	Right
Chicago Cubs	San Diego	0.33	Away team	Wrong
Florida	St. Louis	−0.35	Away team	Right
Milwaukee	Houston	−0.28	Away team	Right
Montreal	Arizona	−0.40	Home team	Wrong
NY Mets	Cincinnati	−0.05	Away team	Right
NY Yankees	Cleveland	−0.14	Home team	Wrong
Oakland	Tampa Bay	0.16	Home team	Right
Philadelphia	San Francisco	−0.08	Away team	Right
Pittsburgh	Los Angeles	−0.02	Home team	Wrong
Seattle	Baltimore	0.26	Away team	Wrong
Texas	Kansas City	0.25	Home team	Right
Toronto	Chicago White Sox	−0.04	Away team	Right

TABLE 10-2 Predictions and Game Results for All Games Played on June 1, 1999

$$Success\ rate\ using\ the\ prediction\ formula = \frac{924}{1616} = 57.2\%$$

How Good Were Our Predictions?

One thing that is a bit surprising is the low value of the success rate. If we guessed at the winners, we would have a success rate of 50 percent. By using information about the team strengths and the home effect, we've raised this success rate to only 57.2 percent. People have tried this same method for predicting games in professional basketball and football and achieved better results. Using the same least-squares method as the one described above, one can get a 63 percent success rate for predicting professional football games and a 69 percent success rate for professional basketball. This tells us that the results of baseball games are pretty uncertain relative to football and basketball.

How could we improve our predictions? Is there other information about the game that one could incorporate, making for a better prediction formula? One obvious piece of information to add would be the quality of the starting pitchers of the two teams. Starting pitchers like Randy Johnson, Greg Maddux, and Pedro Martinez have the potential to dominate a game, and so it would seem that knowledge of the starters should help our predictions.

To investigate this conjecture, we look at the ten best pitchers in each of the National and American Leagues on the basis of ERA, and check the success of our method in predicting the games started by these star pitchers during the months June through September. For each pitcher, Table 10-3 displays the number of games correctly predicted. Also, the table divides the incorrectly predicted games into two groups—the games where the team was predicted to win but lost, and the games where the team was predicted to lose but won. Note from the "Totals" row that our method gave correct predictions in 246 out of 408 games, for a success rate of 60 percent. (This is a little better than our success rate for all games.) Now if the knowledge of the pitcher improved our predictions, we would expect to see more errors where the team won with star pitchers although predicted to lose. Actually, in Table 10-3 we see just the opposite pattern—in 85

Pitcher	Number of Predictions	Correct Predictions	Predicted to Win, but Lost	Predicted to Lose, but Won
P. Martinez	17	12	2	3
D. Cone	19	9	7	3
M. Mussina	18	9	6	3
B. Radke	21	12	7	2
J. Rosado	20	11	5	4
J. Moyer	20	12	3	5
B. Colon	20	12	4	4
M. Sirotka	23	13	4	6
F. Garcia	21	9	6	6
O. Hernandez	21	17	1	3
R. Johnson	23	13	6	4
K. Millwood	21	15	3	3
M. Hampton	21	18	2	1
K. Brown	23	11	8	4
J. Smoltz	19	9	4	6
T. Ritchie	19	9	2	8
C. Schilling	13	8	3	2
G. Maddux	22	15	4	3
J. Lima	25	15	5	5
O. Daal	22	17	3	2
Totals	408	246	85	77

TABLE 10-3   Outcomes of 1999 Predictions in Games Started by the Twenty Best Pitchers

games the teams lost games they were predicted to win, and in 77 games the teams won games they were predicted to lose.

Generally, we were unsuccessful in developing a more useful prediction formula that incorporated the starting pitchers. That doesn't mean that the starting pitchers are not important in determining the game results. Instead, what it most likely means is that the strength of the starting pitchers is already part of the team strengths used in our earlier prediction formula. For example, Pedro Martinez is a great pitcher who helps Boston win some games, but his ability is built into Boston's team strength, which we used in our predictions.

Predicting the Number of McGwire and Sosa Home Runs

Let's shift gears from predicting game results to predicting individual player accomplishments. The 1998 baseball season will forever be remembered as one of the most memorable, largely due to the achievements of Mark McGwire and Sammy Sosa. Before 1998, there were two notable achievements in home-run hitting for a single season—Babe Ruth's 60 home runs in 1927 and Roger Maris's 61 in 1961. McGwire's 70 and Sosa's 66 home runs both shattered Maris's record. Moreover, the two players achieved these marks in dramatic fashion, with McGwire hitting 2 home runs on the final day of the season.

It's no surprise that in the summer of 1998, McGwire and Sosa were the center of media attention, and that every home run hit by either player, especially during August and September, added to the excitement. During the season, everyone wondered: would Mac or Sosa break Maris's home run record, which had stood for 37 years? And if the record was broken, how many home runs would these two sluggers eventually hit?

As we write this chapter in September 1999, with the 1998 season a fond memory, it's not that interesting to talk about predicting 1998 results. So we'll focus instead on one current prediction problem, where the outcome is still not known. It is September 8, 1999, and Sammy Sosa has hit 58 home runs in 535 at-bats. He has 23 games left with 90 at-bats (approximately). Will Sammy break 60? Will he break 70? Will Sammy hit more than McGwire, who currently has hit 54 home runs, with roughly 60 remaining 1999 at-bats?

A Simple Prediction Method

First, let's discuss one simple way of predicting the number of Sosa home runs. This method is probably used by most of the sports sites in the World Wide Web—usatoday.com, espn.com, sportsline.com, and cnnsi.com.

First we compute Sosa's rate of hitting home runs in 1999. He has already hit 58 home runs in 535 at-bats, so his 1999 rate is:

1999 Home-run rate = 58/535 = .108, or 10.8%

Suppose he keeps hitting home runs at the same 10.8 percent rate. So if he has 90 more at-bats, we expect him to hit:

90 (.108) = 9.7, or approximately 10

additional home runs. Since Sosa has already hit 58 home runs, we then predict that his 1999 total will be

1999 Home-run rate = 58 + 10 = 68

So we predict that Sosa will hit a couple of home runs short of the record number 70.

What's Wrong with This Prediction?

There are some problems with this method of prediction. First, although we predict that Sosa will hit 68 home runs, we really have little idea *how likely* it is that he will hit 68 home runs. People who see a prediction that says, "Sosa will hit 68 home runs," will expect that this will happen, and be surprised if Sosa actually hits 66 or 67 or 70 home runs. What we'll learn in this chapter is that there is a lot of uncertainty in prediction. Although a prediction like "68 home runs" may be the most likely possibility based on our knowledge, there is a *greater* probability that Sosa will *not* hit 68 home runs.

Another problem with this prediction is that it is based on a particular set of assumptions, and often fans forget about these assumptions when they see the answer. Here the prediction that Sosa will hit 68 home runs makes the important assumption that Sosa will continue to hit home runs at the same rate as he did in the months April through August. Is this reasonable? Is Sosa really a hitter who hits home runs at a 10 percent rate throughout the season? Or maybe Sosa had an unusually good streak of home-run hitting that stretched through the 1998 season and into the first five months of 1999, and really he isn't as good as this 1998 and 1999 data would indicate.

The point here is that if you gather a group of baseball fans in September 1999, each fan will have his own opinion about the home-run prowess of McGwire and Sosa during the remainder of the 1999 season. One fan might believe that Sosa is really in a groove (with respect to home-run hitting) this year and will continue to stay in this groove for the remainder of the season. Another fan might think that Sosa has been hitting over his head this season and will cool down. And another fan's opinion about Sosa's home-run hitting might be based on the teams and pitchers and ball parks where he will play the 23 remaining 1999 games. Fans will have divergent opinions about the abilities of McGwire and Sosa, and these opinions will result in different predictions about the 1999 home-run totals of these two players.

Can we do anything to reduce some of this uncertainty? In the next section of this chapter, we narrow our focus and describe a simple statistical method for predicting results—for example, the number of home runs by one player in a season, the number of strikeouts by a pitcher, the number of RBIs for a team, and so on. First, we take a look at one player's true home-run rate by means of a probability table. Second, we develop a probability table for the result we are interested in. The probabilities in this table tell us which result is most likely to happen, and reveal that many results besides the most likely one are possible. At the end, we derive a range of possible values, so instead of having to say, "Sosa will hit 68 home runs," we can make a statement like "there is a 90-percent chance that Sosa will hit between 64 and 72 home runs."

A Spinner Model for Home-Run Hitting

Let's first describe a simple probability model for Sosa's home-run hitting for the remainder of the 1999 season. When Sammy comes to bat, we can put the results into three categories:

1. He gets a walk, gets hit by a pitch, or gets a sacrifice—none of which is counted as an official at-bat.

2. He has an official at-bat (a hit or an out) but doesn't hit a home run.

3. He hits a home run.

If we ignore the plate appearances that don't result in an official at-bat, then there are two outcomes—home run or not a home run. Suppose that every time Sammy has an at-bat, he spins a spinner. The spinner has two areas labeled Home Run and Not a Home Run, and the area of the home run region is p. He spins the spinner to bat, and a home run is the result if the spinner lands in the Home Run region.

Note that the chance that Sammy hits a home run on an official at-bat is p for each at-bat. We're assuming that the home-run probability is the same in Wrigley Field, away from Wrigley Field, against Randy Johnson, and against Chad Ogea. Actually, we don't believe that this is true: Sammy's *got* to have a higher probability of hitting a home run against Ogea than Johnson. Still, this model works pretty well in representing the variation of home run-data that we see.

So this model assumes that Sammy's chances of hitting a home run are the same regardless of the pitcher he is facing. And there is a second big assumption here: the chance of Sammy hitting a home run in, say, the twentieth at-bat is not affected at all by his performance in the previous 19 at-bats. We're assuming that he can't have true hot or cold streaks in his hitting. (For a more extensive discussion of this point, see Chapter 5.)

How Many At-Bats?

We don't know for sure how many at-bats Sammy will have in the final 23 games, but we note that he has played in every single Cubs game this season. So it is reasonable to assume that he will play in each one of the remaining 23. Also, in his first 138 games, Sammy averaged 3.9 at-bats. If he continues to get at-bats at this rate, we would expect him to have:

$$23 (3.9) = 89.7, \text{ or about 90 at-bats}$$

So, in our spinner model, we will assume that Sammy will get 90 opportunities to spin the spinner and get home runs.

What If We Knew Sosa's True Home-Run Rate?

To complete our spinner model, we have to know the chance that Sammy will hit a home run on a single at-bat—this is the size of the Home Run area in the spinner.

Let's first assume, hypothetically, that Sammy is a true 10 percent home-run hitter. That is, the chance that he hits a home run on an official at-bat is .1. In our spinner model, the area of the Home Run region would be .1, and Sammy could play the remainder of the 1999 season by spinning this particular spinner 90 times (corresponding to the 90 at-bats). How many home runs would Sammy hit?

Well, a reasonable guess would be 9. Since the chance of hitting a home run is 10 percent, one would expect him to hit $90 (.10) = 9$ home runs in the remainder of the season. But the actual number of home runs he will hit is random or uncertain, and although 9 home runs is pretty likely, there is a good chance that he will hit fewer or more than 9.

There is a well-known formula called the *binomial* that is used for computing such a probability. We will take a close look at it to see if we can determine the probability that Sammy will hit a specific number of home runs for our spinner model with 90 at-bats (spins) and a home-run probability (spinner area) of .1.

Binomial Probabilities

Suppose you have a random experiment that consists of a sequence of trials. On each trial, only one of two things can happen, which we call a success (labeled S) or a failure (F). Assume that the chance of an S on a single trial is p, and that the chance of getting a getting a S or F on a particular trial is not affected by what happens on previous trials. If there are N trials, then the probability of seeing exactly x successes in the experiment is given by the following formula:

$$\Pr(x \text{ successes in } N \text{ trials}) = \binom{N}{x} p^x (1-p)^{N-x}$$

In this formula, the symbol

$$\binom{N}{x}$$

called "N over x," is the number of ways of choosing x items from a larger group of N items. In the example above, the number of trials is $N = 90$, a success is "hitting a home run," the probability $p = .1$, and we are interested in the probability of hitting a particular number of home runs. Table 10-4 displays some of these binomial probabilities.

Looking at Table 10-4, we see that the most likely number of home runs Sammy will hit in the final part of the 1999 season is 9. But the probability that Sammy will hit *exactly* this number is only about 14 percent. That's a small

Home Runs	Probability		
0	0	11	0.101
1	0.001	12	0.074
2	0.004	13	0.049
3	0.012	14	0.030
4	0.030	15	0.017
5	0.057	16	0.009
6	0.089	17	0.004
7	0.119	18	0.002
8	0.137	19	0.001
9	0.139	20	0
10	0.125		

TABLE 10-4 Binomial Probabilities for Number of Home Runs Hit in 90 At-Bats with a Home-Run Probability of .1

probability. Looking further at the table, we see that the probabilities that Sammy hits 7, 8, 9, 10, and 11 home runs are all above 10 percent. The message here is that even if we know Sammy's hitting probability, we aren't too sure about what can happen in Sammy's next 90 at-bats.

What If We Don't Know Sosa's True Home-Run Rate?

But we can't use the results of Table 10-4 to predict the number of home runs Sosa will hit. Why? Well, we don't know for sure the value of Sammy's hitting probability, and it is not reasonable to assume that it is exactly equal to .1.

To get some idea what Sosa's home-run probability in 1999 might be, Table 10-5 shows Sosa's at-bat and home-run data for the previous ten years in the Major Leagues. (For this discussion, we are basing our judgments on Sosa's accomplishments prior to 1999. Later we'll talk about how to change this judgment after seeing the home-run data for the 1999 partial season.) For each year, we have computed Sosa's home-run rate and put those numbers in the last column.

Let's focus on Sosa's home run rates:

.022, .028, .032, .031, .055, .059, .064, .080, .056, .103

Figure 10-1 plots these values as a function of the season. We see an interesting pattern here. For Sosa's first four years, he hit home runs at roughly a 3 percent clip. Over the next five years, he hit home runs at rates between 6 and 8 percent. And in 1998, his rate was over 10 percent!

Year	Team	AB	HR	HR Rate
1989	Texas-ChiW	183	4	0.022
1990	ChiW	532	15	0.028
1991	ChiW	316	10	0.032
1992	ChiC	262	8	0.031
1993	ChiC	598	33	0.055
1994	ChiC	426	25	0.059
1995	ChiC	564	36	0.064
1996	ChiC	498	40	0.080
1997	ChiC	642	36	0.056
1998	ChiC	643	66	0.103

TABLE 10-5   Number of At-Bats and Home Runs for Sosa in His Major League Seasons Prior to 1999

Based on Sosa's career home-run statistics, we have to make some judgment about the value of his home-run probability for the 1999 season. From his stats, it appears that Sammy has matured considerably as a home-run hitter, and the statistics for the last few years are probably the ones that are most representative of his current ability. But we also have to realize that the home run rates in Figure 10-1 are really not hitting probabilities, but observed home run rates. Maybe Sammy was not a real 10 percent home-run hitter in 1998, but was lucky and had a good year.

After some reflection, we realize that we're pretty uncertain about Sammy's home-run probability for 1999. Based on his 1998 season, we believe that he has the potential to be a "great" home-run hitter where his home-run probability is 10 percent or higher. (We consider 10 percent a useful reference point, since it corresponds to the observed home-run percentage of Babe Ruth in his prime.) In addition, based on the pattern of home-run hitting shown in Figure 10-1, he appears to be fully matured, or at least close to fully mature, as a home-run hitter. However, we can't forget the relatively small home-run rates that he achieved just a few years ago. Based on these beliefs, we constructed the bar chart in Figure 10-2 to graphically represent his 1999 home-run probability. Remember again that these beliefs are based only on Sammy's accomplishments for the ten-year period 1989–1998; for the moment, we're ignoring the 1999 data.

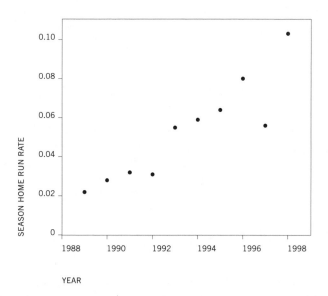

FIGURE 10-1   Sosa's home run rates for the first ten years in the Major Leagues.

Looking at the probability graph, we see that the values of .09 and .10 are each assigned the largest probabilities. This means we think it's most likely that Sosa will have a 9 or 10 percent home-run probability in 1999. However, as the figure indicates, we're not sure that his home-run probability is 9 to 10 percent, and we think it is possible that his probability can be as small as 5 percent or as large as 13 percent. Actually, our beliefs about Sosa's chance of hitting a home run are pretty vague. Before the 1999 season begins, we're not sure if he will continue to hit home runs at his 1998 rate, or revert back to his pattern of hitting home runs in the earlier seasons.[1]

Revising Our Beliefs about Sosa's Home-Run Probability

We've discussed how to construct a probability table that reflects our beliefs about Sosa's home-run probability in 1999. Now we watch Sosa's batting performance in the first five months of the season, and we observe 58 home runs in 535 at-bats. This is a pretty impressive performance, and we're more confident that Sammy's home-run probability in 1999 is high. So we want to revise the probabilities we displayed in Figure 10-2 in light of this new batting data. Fortunately, there is a simple formula, called *Bayes' Rule,* that tells one how to change one's probabilities when given new data.

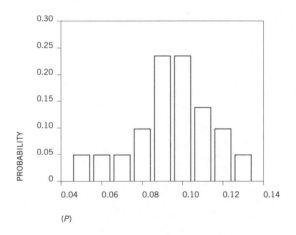

FIGURE 10-2   Graph of probabilities for Sosa's true home-run rate.

1 As this book is going to press, at the beginning of the 2001 season, we have seen Sosa's home run statistics for the 2000 season, and we are more certain that he is a hitter with a 8–12-percent home-run probability.

We'll use Bayes' Rule to update the probabilities. We are interested in learn-ing about a batter's home-run probability p, and our initial beliefs about p are represented by means of a prior probability distribution. After we observe some data, then our new, or posterior, probability distribution for p is given by this formula:

$$\text{Pr}(p \text{ given Data}) = \text{Pr}(p)\, \text{Pr}(\text{Data given } p)\, / \, c$$

where c is the probability of observing Data based on our initial opinion. The value of c ensures that the probabilities add up to 1.

In our setting, our DATA is "58 HR in 535 AB." To find the new probability that Sosa's hitting probability is 10 percent, we compute the product:

$$\text{Pr}(p = .1 \text{ given } 58 \text{ HR in } 535 \text{ AB}) = \text{Pr}(p = .1)\, \text{Pr}(58 \text{ HR in } 535 \text{ AB given } p = .1)$$

$$= (.2353)\,(.0451)$$

$$= .01062$$

In the formula, $\text{Pr}(p = .1)$ is our initial probability that Sosa is a 10 percent hit-ter and $\text{Pr}(58 \text{ HR in } 535 \text{ AB given } p = .1)$ is the binomial probability that Sosa gets 58 home runs in 535 if he really is a 10-percent home-run hitter.

In Table 10-6, the product $[\text{Pr}(p)\, \text{Pr}(\text{Data given } p)]$ is found for each of the pos-sible home-run probabilities for Sosa. In this table, each of the entries in the Product column is divided by c, the sum of the products, to get the new (poste-rior) probabilities in the last column.

The probabilities in the last column of Table 10-6, $\text{Pr}(p \text{ given Data})$, reflect our beliefs about Sosa's home-run probabilities after seeing his 1999 data. Remember that, before seeing Sosa perform in 1999, we thought he was a 9 or 10 percent home-run hitter—with a small chance of being either a 5 to 6 percent or an (unimaginably great) 13 percent home-run hitter. After seeing Sammy's per-formance in the first five months of 1999, we see that the values of p in the set {.09, .10, .11, .12} have most of the probability, which means we're pretty confi-dent that Sammy's home-run probability is in the 9- to 12-percent range.

One Prediction

We are finally ready to predict the number of home runs Sosa will hit in the remainder of the 1999 season. Recall that if we really knew Sosa's 1999 home-run probability p, then we could compute the probability that he would hit a par-ticular number of home runs using a binomial formula. We don't know the value

p	Pr(p)	Pr(Data given p)	Product	Pr(p given Data)
0.05	0.0490	0	0	0
0.06	0.0490	0	0	0
0.07	0.0490	0.0003	0.00001	0.0005
0.08	0.0980	0.0039	0.00038	0.0136
0.09	0.2353	0.0195	0.00458	0.1647
0.10	0.2353	0.0451	0.01062	0.3816
0.11	0.1373	0.0550	0.00755	0.2714
0.12	0.0980	0.0390	0.00383	0.1375
0.13	0.0490	0.0174	0.00085	0.0306
	1		0.02782	1

TABLE 10-6   Bayes' Rule Computations to Obtain Updated Beliefs about Sosa's New Home-Run Probabilities p

of the home-run probability p, but our beliefs about this probability are described by the probabilities shown in Table 10-6.

The probability that Sosa hits a given number of home runs, say 10, is given by the formula:

Pr(10 HR) = sum of [Pr(HR prob. is p) × Pr(10 HR if the HR prob. is p)]

for all possible values.

We use Table 10-7 to illustrate how we compute the probability that Sosa hits 10 additional home runs in the 1999 season. The first column lists the possible values of the hitting probability, the second column lists the corresponding probabilities from Table 10-6. The third column lists the probability that Sosa gets 10 home runs for each probability value. To get the probability of 10 home runs, we multiply, for each row, the values in the second and third columns— and the products are placed in the fourth column. The sum of the products is the probability of interest.

Suppose that we repeat this calculation for all possible home-run numbers. Table 10-8 displays the following probability table for the number of additional home runs Sammy will hit in 1999.

From this probability table, we can make the following predictions. Remember, Sosa has already hit 58 home runs, and this table tells us how many additional home runs he will hit in his future 90 at-bats. On September 8, 1999:

- It is most likely that Sosa will hit 9 more HRs (for a total of 67), but the chance of this happening is only about 13 percent.

p	Pr(HR prob is p)	Pr(10 HR if the HR prob is p)	Product
0.05	0	0.0092	0
0.06	0	0.0245	0
0.07	0.0005	0.0486	0
0.08	0.0136	0.0779	0.0011
0.09	0.1647	0.1055	0.0174
0.10	0.3816	0.1250	0.0477
0.11	0.2714	0.1326	0.0360
0.12	0.1375	0.1282	0.0176
0.13	0.0306	0.1144	0.0035

$Pr = 0.1233$

TABLE 10-7 Illustration of the Computation of the Probability that Sosa Hits Ten Additional Home Runs

Home Runs	Probability		
0	0	11	0.105
1	0.001	12	0.082
2	0.004	13	0.060
3	0.011	14	0.040
4	0.027	15	0.025
5	0.050	16	0.014
6	0.080	17	0.008
7	0.106	18	0.004
8	0.125	19	0.002
9	0.131	20	0.001
10	0.123	21	0

TABLE 10-8 Probability Table for the Number of Home Runs Sosa Will Hit in the Remainder of 1999

- There is a high probability (.9025) that Sosa will hit between 5 and 14 additional home runs (for 1999 totals between 63 and 72). It would be a bit surprising if Sammy hit fewer than 5 or more than 14 home runs in the remainder of the season.

- The chance, at this point in the season, that Sammy will break the record is the chance that he will hit 13 or more home runs, which is:

Pr(Sosa hits 71 or more) = .154

Many Predictions

We have focused on predicting Sosa's home-run totals at a particular point of time during the 1999 season. But there is nothing special about September 8—this prediction procedure can be used at any point in time during the season. Our beliefs about Sammy's hitting probability are based on our knowledge about Sammy prior to 1999 and any home-run data we've observed in 1999 up to that particular point in time.

Figure 10-3 shows our predictions for Sosa's 1999 home-run total. After each game that Sosa played in that season, a vertical line shows the limits of a 90-percent prediction interval for the 1999 home-run total. Before the season began (at Game 0), our prediction interval is seen from the graph to be (33, 81). This interval seems pretty wide, but we had little clue in early April how many home runs Sosa would hit. We were not sure what Sosa's true home-run rate (the value of p) would be in the 1999 season, and he hadn't yet hit any home runs in 1999. As the season proceeds and Sammy is hitting home runs, we see from the graph that the length of the prediction interval shortens consider-

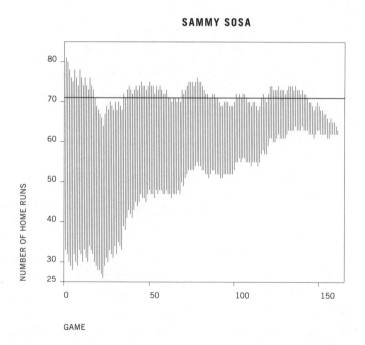

FIGURE 10-3   90-percent prediction intervals for Sosa's 1999 home-run total after each game played in the 1999 season.

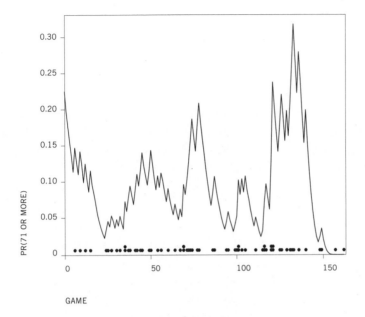

PR(71 OR MORE)

GAME

FIGURE 10-4   Graph of predictive probability that Sosa will break the home-run record of
70 for each game of the 1999 season. The dots along the horizontal axis
show the games in which Sosa hit his home runs.

ably—and eventually, near the end of the season, we are pretty sure about the
final total. (Obviously, when Sosa finishes the season, we know exactly how
many home runs he hits in 1999.)

In this season, many fans were wondering if Sosa would break the season
home-run record of 70. In Figure 10-3, a horizontal line has been drawn at 71
home runs, which represents a new record. Note that for much of the season,
our 90 percent prediction interval covers 71, which indicates that Sosa had a
significant probability of breaking the record. This point is reinforced in
Figure 10-4, which graphs the predictive probability that Sosa will break the
home-run record after each game of the 1999 season. (The dots at the bottom
of the graph show when Sosa hit his home runs.) Note that whenever Sosa hit
one or more home runs during a game, the probability that he breaks the
record jumps up. This increase in the probability has two explanations. First,
since he has hit home runs, he is closer to the record of 71. Also, the fact that
he has hit home runs increases the likelihood that his home-run hitting probability (p) is large.

Finally note that the number of home runs that Sosa actually hit in the 1999
season, 63, is included in most all of the prediction intervals that we constructed

that season. Although Sosa slumped a little at the end of the season, his final total of 63 was consistent with the predictions that we made using our model.

Let's compare Sosa with Mark McGwire. The same method we applied above can also be used to predict McGwire's 1999 home-run total. One difference in the analysis of McGwire is that our initial beliefs (before the season started) about Mark's 1999 true home-run probability, p, are notably different from our beliefs about Sosa's. Figure 10-5 shows the probabilities we used. There is a lot more evidence from past seasons that McGwire hits home runs at a high rate, so we place a high probability on the likelihood that he will hit home runs at a 10- to 13-percent clip. Another difference in our predictions is that we assume that McGwire averages only 3.2 official at-bats per game. (McGwire generally walks more than Sosa, resulting in fewer official at-bats.)

Figure 10-6 shows our 90-percent prediction intervals for Mark McGwire. The pattern in these predictions is very different from the pattern in Sosa's graph (Figure 10-3). Before the season started, we predicted that McGwire would hit 60 home runs with a prediction interval of (41, 81). McGwire started the season slowly, so the predictions dropped off substantially. In fact, for most of the season, we predicted that Mark's 1999 home-run total would be in the mid 50s, and he had essentially no chance of breaking his 1998 record of 70. But McGwire's home-run hitting picked up toward the end of the season, with a final flurry that gave him the 1999 crown. Nonetheless, during most of the season, it was Sosa who had the greater chance of setting a new record.

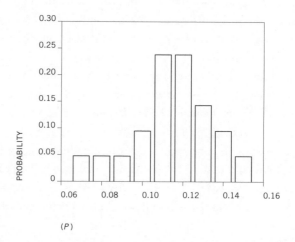

FIGURE 10-5   Graph of probabilities for McGwire's true home-run rate.

MARK MCGWIRE

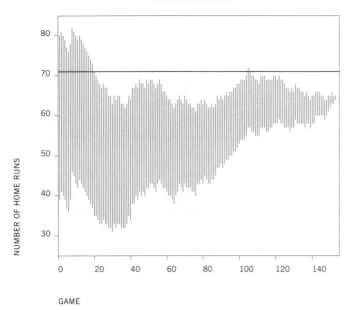

FIGURE 10-6   90-percent prediction intervals for McGwire's 1999 home-run total after each game played in the 1999 season.

Predicting Career Statistics

The previous discussion focused on predicting the number of home runs at some time during a season. But Sammy Sosa is currently only 32 years old, and he will likely play in the major leagues for a good number of years. How many home runs will he hit in his career? Will he break the 500, 600, 700 career home-run marks? And what's the chance he will eventually break McGwire's season home-run record of 70?

This is a more difficult prediction problem, since there are more unknowns. Earlier we concentrated on learning about Sosa's home-run probability p only during the 1999 season. Now we have to think about Sosa's home-run probabilities for each of the remaining seasons of his career. We don't know how many years Sosa has left in his baseball career. And, even if we knew that he would play for, say, eight additional seasons, we don't know exactly how many at-bats he will have.

Nonetheless, in this section we will describe one statistical model for predicting Sosa's career home runs. We will first talk about a model for Sosa's home-run probabilities. Next, we'll discuss the number of opportunities (years and at-bats) that Sosa may have to hit home runs, then use this model to make our predictions.

Sosa's Home-Run Probabilities

In our earlier discussion, we estimated Sosa's 1999 home-run probability p using two types of information: our beliefs about his home-run probability prior to 1999, and his hitting data during the 1999 season. How can we learn about Sosa's home-run probabilities in the years 2000 and beyond, when we haven't yet observed any data for these years?

Obviously we can't look into the future, but it is reasonable to believe that Sosa's pattern of hitting home runs over his career will be similar to the hitting pattern of other sluggers in history. Figure 10-7 plots the home-run hitting rates (HR divided by AB) against the player's age for the nine greatest career home-run hitters in baseball history. Smooth curves are drawn over each of the graphs to show the basic patterns in the rates. Looking at the graphs, we note that there

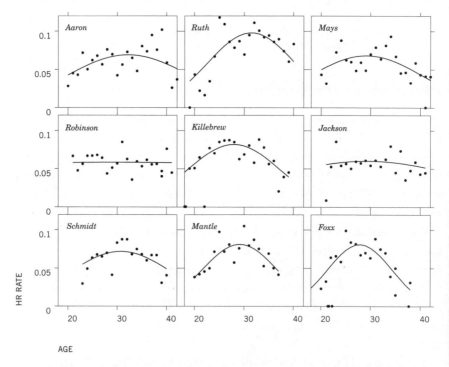

FIGURE 10-7   Observed home-run rates and smooth fitted curves for the career performances for nine great home-run hitters.

are significant differences between these ten great players in terms of their home run rates. But we notice a general pattern, as shown in Figure 10-8.

Generally, the curve shows us, a home-run hitter improves (matures) in his first few years, reaching a peak near the age of 30. After that peak, the home-run hitter tends to decline. Although this pattern seems to hold for most players, there are differences in the peak age and the degree of maturation and decline between players. For example, Henry Aaron peaked relatively late, then declined relatively slowly. In contrast, Mickey Mantle peaked at an earlier age but showed dramatic declines later in his career.

Using Sosa's home-run data through the 1999 season and the career statistics of the fifty greatest home-run hitters, we can learn about Sosa's home-run probabilities for the remainder of his career. We assume that Sosa's home-run probabilities for his career will follow the basic shape, and assume that his career pattern of home-run probabilities will be similar to the pattern of the other 50 home-run hitters.

We don't know exactly what Sosa's "true" home-run rates will be, but we can generate sets of home-run rates, as shown in Figure 10-9, that we think are reasonable based on our model. The dark solid line represents our best guess at what Sosa's home-run probabilities will look like over the years. We think that he will peak at age 31 (in the year 2000) at the value .09, then the probability will decrease to a value of .04 when Sosa is 40 years old. But this figure shows that, even though we have a best guess at his home-run probabilities, it is possible that they will deviate a bit from this best guess. Specifically, it is possible that Sosa will peak at a later age and show faster or slower rates of decline.

How Long and How Many At-Bats?

Now that we have a handle on Sosa's home-run probabilities, we have to next decide how many years Sosa has in his career. We really don't know much about this. Sosa appears to be a well-conditioned athlete and likely has many productive years ahead of him. But many things, like injuries, could have an impact on the length of his career. We will first assume that Sosa could play until the ripe old age of 41, but since this is an important assumption, we will present predictions assuming that Sosa does retire at an earlier age.

FIGURE 10-8   Pattern of home-run rates for great home-run hitters.

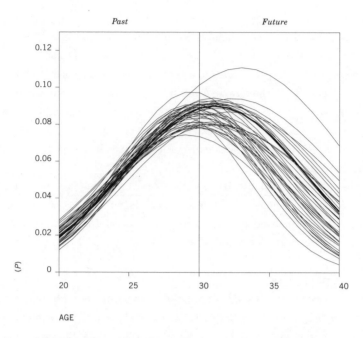

FIGURE 10-9   Plausible graphs of Sosa's home-run hitting probabilities based on the model.

Given his past playing performance, it is reasonable to think that Sosa will continue to play regularly and have a large number of at-bats for a number of years. However, for most players, the number of batting opportunities (at-bats) does decrease by 5 to 10 percent as the player approaches the twilight of his career. So we assume that Sosa's at-bats in the coming years will look something like this:

Sosa's Age (Years)	Expected At-Bats
31, 32, 33	620
34, 35, 36	590
37, 38, 39	560
40, 41	500

Of course, a lot could happen in Sosa's career that will cause him to have fewer at-bats during particular years, but these values seem consistent with the pattern of at-bats for other home-run sluggers in history.

Making the Predictions

Now we're ready to make our predictions. What we do is perform a large number of simulations for the remainder of Sosa's career using the probability model we have constructed. We first simulate a set of home run probabilities from the model described earlier. Each of these probabilities defines a random spinner where the area of the Home Run region of the spinner corresponds to the probability. In the particular simulation illustrated in Figure 10-10, Sosa's home-run probability at ages 31–34 is 10 percent, his home-run probability at ages 35–36 is 9 percent, his probability at age 37 is 8 percent, and so on. Then we use the random spinners to simulate home-run results using the at-bat numbers given above.

Table 10-9 summarizes the results of doing our simulation for a total of 1000 Sosa careers. Since the results depend heavily on how long Sosa remains active in the major leagues, we give results assuming that Sosa plays until particular ages. In each case, the table gives a "best guess" at Sosa's number of career home runs, and the chances that he will break the home-run milestones of 500, 600, and 700. Finally, we give the chance that Sosa will sometime break the single-season home-run record of 70.

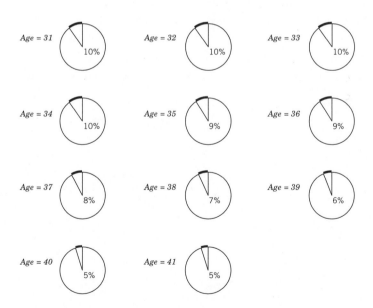

FIGURE 10-10 Set of random spinners corresponding to a particular simulation of Sosa's home-run probabilities.

Plays until Age	32	34	36	38	40
Best Guess at Career HR	438	527	597	644	673
Pr(500+)	0	0.81	0.98	0.99	0.99
Pr(600+)	0	0.01	0.48	0.73	0.79
Pr(700+)	0	0	0.03	0.22	0.37
Pr(71+ season)	0.03	0.06	0.06	0.06	0.06

TABLE 10-9   Results of Simulation of Sosa's Career Using Our Probability Model

Several interesting things can be learned from Table 10-9. First, the number of career home runs depends basically on how long Sosa will remain an active player. Hank Aaron, with 755, currently owns the career home-run total. Sosa has a good chance of breaking 700 career home runs only if he stays active until at least 40. Also, note that Sosa's chance of breaking McGwire's season home-run mark is small—only about 6 percent. Why? Well, we are assuming that Sosa is currently at or near his peak, and his home-run ability will start decreasing-meaning it will be less likely that he will break the mark later in his career.

Then again, Sammy is looking good, looking healthy, looking over pitches with that sparkle in his eye, so we would not bet against him. Nobody's going to make tons of money following the prediction rules described in this chapter. But hopefully we are now more aware of the great amount of uncertainty in prediction. Predictions like "Sosa will hit 60 home runs this year" have a lot of vagueness connected with them, and anyone who tells you that he or she can make more accurate predictions than the ones described here is misinformed, or lying.

DID THE BEST Team Win?

CHAPTER 11

Major League Baseball is currently a competition between 30 teams, 16 in the National League and 14 in the American. Each league is divided into three divisions—East, Central, and West. Each team plays a 162-game schedule during the regular season, playing most of its games against teams in its own league. We say "most" because beginning with the 1997 season, MLB has experimented with inter-league play, where each team from the National League plays approximately 15 games against selected opponents from the American League.

At the end of the regular 162-game season, the four best teams from each league continue playing games in a "post-season," with the goal to win the championship. The teams with the highest winning percentage in each of the three divisions compete together with a "wild-card" team, which has the best winning percentage among all teams who are not division winners. The eight teams go through a three-tier playoff system to decide the championship of baseball. First, the four teams in each league compete in two playoffs—each pair of teams plays a "best of 5" competition to decide the final two teams of each league. Next, each pair of teams in each league play a "best of 7" competition to decide the winner of each league. Last, the winners of the American and National Leagues pennants play a "best of 7" competition called the World Series to decide the champion.

The Big Question

The winner of the World Series is declared the "best team in baseball" and is immortalized as one of the premier teams in the history of the game. Each of the players on the championship team receives a special World Series ring, and there are substantial bonuses paid to each player, the manager, and the coaches for their achievement. However, after all of the games have been played and the winner declared, a natural question to ask is: "Did the best team actually win the championship?" In 1997, the Florida Marlins defeated the Cleveland Indians to win the World Series. This particular contest was very close—the series was not decided until the last extra inning of the seventh game. Many Cleveland Indians fans thought the Indians should have won the series, focusing on several pivotal plays during the series that greatly influenced the outcome. If a pitcher had not made one particular poor pitch, the Cleveland fans felt, or if a player had executed cleanly instead of misplayed in one particular defensive mishap, the Indians would have won the championship. These fans may feel that the Indians were indeed the better team, but, due to some unfortunate circumstances or bad luck, their team lost the series.

Ability and Performance

What does it mean to be the "best team" in baseball? Since the Marlins won the World Series, most people would refer to the Marlins as the best team in 1997. After all, they did win the championship. They were "best" in the sense that they performed the best in the series of playoffs following the regular season. But that's not what we're talking about. What we mean is, "Did the team with the greatest ability win the World Series?" Did the team with the best players, that is, the most talented players, win the championship? Here we are again making the important distinction between ability and performance. The Indians may have had a more talented team in 1997, but they may not have performed to the best of their abilities. Alternatively, the Marlins may have had less overall ability than the Indians, but they could have performed particularly well during the 7-game series to win the championship. In other words, chance could have played a major role in the World Series.

Looking over the recent history of Major League Baseball, we see teams that performed in a relatively average fashion during the regular season but somehow won the World Series. A good recent illustration of this phenomenon is the Minnesota Twins in 1987. The Twins that year finished with a season record of

85 wins and 77 losses, winning just 52 percent of their games. The team scored 786 runs during the season and allowed 806, so it is surprising that they even had a winning season. Their team batting average was a lackluster .261, which was lower than the American League average of .265, and their slugging average of .430 was only slightly higher than the league average of .425. You may be thinking that they must have had good pitching. Well, their team ERA was a weak 4.63, which was higher (worse) than the league ERA of 4.46.

This array of less-than-stellar stats may lead you to ask, "Did they do anything well?" The answer is yes, the Twins had the fewest number of fielding errors in the major leagues. More importantly, they played well in their home ball park. Their record during the regular season was 56–25 (a winning percentage of .691) at the Metrodome and 29–52 (a winning percentage of .358) on the road. The Twins continued this pattern of winning during the post season—they won all six games played at the Metrodome during the American League Championship and World Series. Nonetheless, although the Twins won the World Series in 1987, it would be difficult to argue, on the basis of their team statistics, that they were the best team in baseball that particular year.

Similarly, if one looks at team statistics, one can question if the Florida Marlins, the winner of the 1997 World Series, was the best team ten years later. Let's compare the Marlins with the Atlanta Braves, a team that played in the same National League division. Table 11-1 lists a number of statistics for the two teams that year. We see that Atlanta had a higher winning percentage, scored more runs, allowed fewer runs ("OR" stands for Opponents' Runs), had a higher team batting average and slugging percentage, and had a lower earned-run average. When Florida defeated Atlanta in the National League Championship that season, a number of explanations were offered, and some argued seriously that Florida was the better team. But based on the regular season statistics, it should be pretty clear that Atlanta was the superior team and was more deserving than Florida of a World Series championship.

Note that we observe a baseball team's performance throughout a season, but we never know for sure, even after the last game of the year, exactly how tal-

	Wins	Losses	Win %	R	OR	AVG	SLG	ERA
Florida Marlins	92	70	0.568	740	669	0.259	0.395	3.83
Atlanta Braves	101	61	0.623	791	581	0.270	0.426	3.18

TABLE 11-1   Team Statistics for Atlanta and Florida for the 1997 Baseball Season

ented the team is. The Marlins were better than the Indians in the sense that they performed better during the World Series. An Indians' fan may argue that the Indians were a better team than the Marlins—he's saying that the Indians had more ability. This statement can't be refuted by a Marlins fan, since he or she really doesn't know which team had more talent.

Describing a Team's Ability

Since a team's ability or talent is an abstract quantity that is unknown, it is helpful to use a number to describe it. We will denote the talent of a team by the letter t. If t is equal to 0, then we can think of the team as having average talent. A negative value, say $t = -.4$, will correspond to a team with below-average ability, and a positive value (like $t = +1.2$) will correspond to a team whose talent is in the upper half of all major-league teams (see Figure 11-1). If a team's talent is a positive number, then we expect the team to win more than half of its games, although we will see that the team may not win more than half of its games during a 162-game season.

Each team in the major leagues can be assigned a number t that corresponds to its talent. Since there are currently 30 major-league teams, there exist 30 numbers $t_1, \ldots, t_{30}$ that correspond to their abilities. The problem is that we never know exactly the values of these talents; in fact, we could know them to a reasonable degree of certainty only if the teams were able to play millions of games during the season. Clearly that's impossible, since a baseball season is scheduled over a six-month period, so we view these abilities as unknown hypothetical quantities.

Describing a Team's Performance

So are we stuck? Since we will never know the abilities of these major-league teams, can we go no further? No. We get information about the teams' abilities by observing their performances during a 162-game season. Each team gets an opportunity to play all the teams in its respective league, and they win and lose

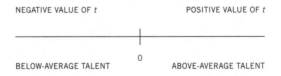

FIGURE 11-1   Interpreting the talent (t) of a team.

games. At the end of the 162-game season, we observe winning fractions for all the teams. We'll use the letter p to denote a team's proportion of wins for the season. So, for example, if the Baltimore Orioles win 90 and lose 72 games during the season, the value of p for Baltimore is $90/(90+72) = .56$. We observe these winning fractions for all 30 major-league teams and denote them $p_1, \ldots, p_{30}$. These numbers are simply the winning percentages reported in the team standings after the last day of the regular season. (We should apologize for the change in notation—p represented an ability in earlier chapters.)

The primary goal of this chapter is to show how the baseball teams' abilities, as measured by the talent numbers, are linked to their season performances, which are described by the observed winning fractions. We'll first look at baseball teams' winning percentages since the beginning of professional baseball (1871). This investigation will show that baseball teams appear to be similar in their performances over time. Then we'll look at a few simple models which relate the teams' abilities to their season performances. Once we have found a simple model which seems to describe baseball competition reasonably well, we'll use the model to relate teams' abilities with their season performances. To whet your appetite, we'll address the following questions (among others):

1. How does baseball competition in 1997 relate to competition during the 1920s? Were teams more similar in ability back then?

2. How does a team with average ability perform during the regular season? Can this average team ever win the World Series? On the other hand, can this average team finish last in their division?

3. Suppose a team like the Marlins wins the World Series and is declared the best team in baseball. What's the chance that the Marlins were indeed a team with great ability? What's the chance that the Marlins were an average team? What's the chance that there was a team in the major leagues that year with greater ability than the Marlins?

Team Performance: 1871 to the Present

How have baseball teams performed in the past? From the first days of professional baseball in 1871, records have been kept of the winning percentages for all teams. Figure 11-2 plots all of the team winning percentages against the season year. There are a number of interesting features that one can see from this graph.

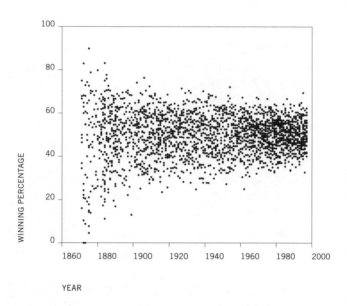

FIGURE 11-2   Scatterplot of team winning percentages and season year.

- There was a large spread in the winning fractions in the early years of baseball—from 1871 to the late 1800s. Some teams won 10 percent or fewer of their games, and other teams won over 80 percent of theirs.

- Generally, as we look from left to right in the graph (earlier to later years), we see that the spread in the winning fractions appears to get smaller. One way to notice this change is to focus on the teams winning between 60 and 70 percent of their games. There have always been teams that have performed this well, from 1871 to the present. However, it seems that the number of teams in this category is decreasing relative to the number of average teams, which by our definition win between 40 and 60 percent of their games. This same comment is true for weak teams that win only 30 to 40 percent of their games. The fraction of teams that perform this poorly has appeared to decrease over time.

- Reinforcing the previous comment, note that the winning percentages in the early years, say 1880–1900, appear to be uniformly spread out from 30 to 70 percent. In contrast, practically all of the winning percentages in the last few years have been located in the 40- to 60-percent range. Sure, occasionally there are poor or weak teams that win 30 percent and 70 percent of the time, respectively, in recent years. But such occurrences are pretty rare, and the trend seems to be toward more "average" performance.

Explanations for the Winning Percentages

What are possible explanations for the patterns we note in Figure 11-2? Have there been some changes to the structure of baseball competition that might account for them?

Let's first look at the number of games played in a season for all of these teams. Figure 11-3 plots the number of games played for all teams as a function of year. From this graph, we see that in the early years of baseball, seasons were relatively short. In the beginning, seasons were only 20 games long, but by 1900 there were 154 games in a season. From the turn of the last century to 1960, the number of games played averaged about 150. In 1961 (the historic year when Roger Maris hit 61 home runs), the number of games increased to 162, which is the length of the current season.

This graph partially helps to explain the spreads in winning percentages that we saw in Figure 11-2. In the early years, the large spread in winning percentages is partly due to the fact that the seasons were short. Because of the varying

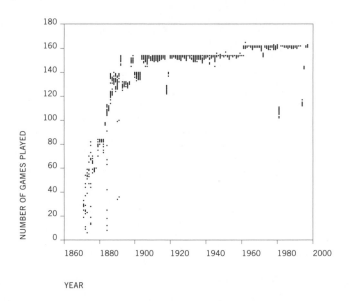

FIGURE 11-3 Number of games played by professional teams plotted against year.

season lengths in the pre-1900 years, it is harder to compare the winning percentages in these early years with the present day. A team with a winning percentage of 80 percent in 1880 wasn't necessarily better than the great teams of the 1990s. This 120-year-old winning rate of 80 percent may reflect the basic truth (from statistics) that it is easier to win 80 percent of only 60 games than 80 percent of 162 games. Because of the varying season lengths in these early years of baseball, we will focus our analysis on the teams in recent years, where the seasons were generally from 150 to 162 games long.

So one change in baseball competition over the years is in the length of the season. What about the number of professional teams in baseball? Figure 11-4 plots the number of teams against the year. We see that in the early years there were many changes in the basic competitive structure of the sport. In some years there were fewer than 10 teams in the major leagues, and in one year there were over 30 professional teams. But starting with 1900, the number of teams stabilized. In fact, in the 60 years from 1901–1960, there were generally 16 teams—8 in the National League and 8 in the American.[1] Then, starting in 1961, professional baseball embarked on its modern expansion. Two new teams were added in 1961, two in 1962, four in 1969, two in 1977, and two in 1993. This expansion may have had an impact on the winning percentages observed in Figure 11-2. As

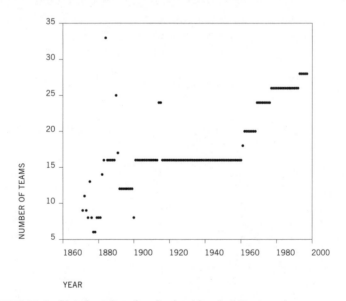

FIGURE 11-4   Plot of number of professional baseball teams and year.

[1] A third major league, the Federal League, with 8 teams, existed briefly in 1914–1915.

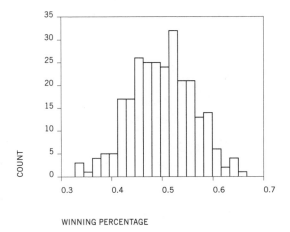

FIGURE 11-5 Histogram of winning percentages for ten recent years.

one adds new teams to baseball, one can speculate that the pool of available ballplayers is spread out over all of the teams, which might make the teams more similar in ability. This similarity in ability is reflected in the small spread in the winning percentages in the last 20 years.

A Normal Curve Model

Let's focus on the group of winning percentages during the 12 years 1986–1997, excluding the years 1994 and 1995. (A baseball strike in 1994 and 1995 resulted in canceled games and significantly shortened seasons.) For the remaining 10 seasons, the number of games for all teams was pretty constant (from 160 to 162).

Figure 11-5 displays a histogram of all of the teams' winning percentages for this modern ten-year period. We see that that the curve of winning percentages is bell-shaped, with most teams winning between 45 and 55 percent of their games. The distribution is symmetric about the value of .5, which corresponds to an average team that is winning half of its games. It seems pretty uncommon during these years to have percentages smaller than 40 or larger than 60. Since winning over 60 percent of the games is a relatively rare event, the few that reach or exceed that number can be viewed as outstanding.

Since the distribution of winning percentages is mound-shaped, one can model this distribution by using a smooth curve—the so-called *normal curve* frequently used in statistics. As discussed in an earlier chapter, a normal curve is bell-shaped and is described by two numbers, a mean M and a standard devia-

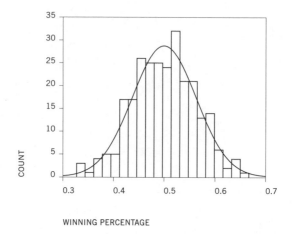

WINNING PERCENTAGE

FIGURE 11-6   Histogram of winning percentages for ten recent years with a normal curve placed on top.

tion S. For a normal curve, the mean M is the middle or most common value of the curve. Here a good choice for the mean is $M = .5$, which corresponds to the winning fraction for an average team. The standard deviation S is a positive number that reflects the spread of this curve. One can choose a value of S by computing the standard deviation of the winning fractions of the modern teams. Here the standard deviation turns out to be .0626. So a normal curve with mean $M = .5$ and standard deviation $S = .0626$ appears to be a reasonable match to this set of winning percentages. To check this out, Figure 11-6 shows the histogram of winning percentages with the normal curve drawn on top. It seems to be a good fit.

This normal curve provides a convenient description of the performances of modern baseball teams during a 162-game season. If the data follow a bell-shaped curve, then roughly 68 percent of the winning fractions will fall within 1 standard deviation of the mean, and 95 percent will fall within 2 standard deviations of the mean. If we apply these rules in this setting, we find the following:

- 68 percent of the winning fractions will fall between [.5 − .0626] and [.5 + .0626], or .44 and .56.

- 95 percent of the winning fractions will fall between [.5 − 2(.0626)] and [.5 + 2(.0626)], or .37 and .64.

These statements help us to understand season performances of teams. For a team to win only 35 percent of their games during a season is a bit unusual, since only 5 percent of all winning percentages are smaller than 37 or larger

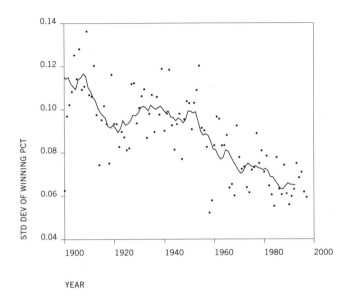

FIGURE 11-7   Plot of standard deviations of season winning fractions for different years. A smooth moving-average curve is placed on top.

than 64. Also, these statements reinforce the parity of baseball. About two-thirds of all teams have winning percentages between 44 and 56. One can interpret this statement as saying that most teams have season performances that are close to average.

Team Performances over Time (Revisited)

Season performances of all baseball teams are generally bell-shaped, as illustrated in Table 11-6, and the standard deviation gives a useful measure of the similarity of the teams. Let's return to Figure 11-2, which plotted all of the winning fractions against the season year. We now have a measure, the standard deviation, that can be used to describe the spread of performances for each year. Figure 11-7 graphs the standard deviations of the season winning fractions against the season number. We focus only on the seasons since 1900, since that is the point from which the lengths of the seasons are pretty constant. There is a lot of scatter in the graph shown in Figure 11-7. The standard deviation can be influenced heavily by a few extreme values, which would correspond to teams with unusually good or poor seasons. But there is also a clear pattern in this graph, which is visible in the smooth curve through the points as shown in

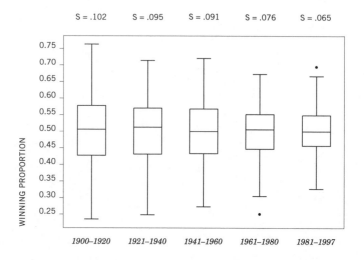

S = .102   S = .095   S = .091   S = .076   S = .065

FIGURE 11-8   Boxplots of season winning proportions for teams of different eras. The standard deviation of each group of proportions is indicated above each boxplot.

Figure 11-7. This smooth curve is found for a given year, say 1980, by computing the average of the 10 years that are close to that year (1975–1984).

Remember, the standard deviation measures the disparity of the group of winning fractions. In the early 1900s, teams had a greater disparity in performances (winning percentages), which is reflected in large standard deviations. The team performances became more similar until about 1920, when they seemed to get more divergent. From the 1950s to the present day, the standard deviations of the winning fractions have been decreasing again, which means that the teams are becoming more alike in their season performances. To reinforce this point, Figure 11-8 shows boxplots of the winning fractions for teams from different time periods. The leftmost boxplot displays the fractions for the teams from 1900–1920, the next boxplot shows the fractions for the 1921–1940 teams, and so on. The associated standard deviation for each group of winning fractions is shown at the top of the figure. We see that the boxplot with the smallest spread corresponds to the 1981–1997 values, which results in the smallest standard deviation value.

One important finding in this analysis is that the current major-league teams appear to be more similar in ability than teams in any time in baseball history. This conclusion will have a significant effect on our exploration into the relationship between teams' abilities and their season performances.

A Mediocrity Model for Abilities

Let's return to our basic question. What is the relationship between the abilities of the current major-league teams and the performances of these teams during the 162-game season? We looked at the season performances, that is the winning percentages, of baseball teams and noted that modern teams appear to be relatively similar in their abilities.

In general, a *model* relates parameters or characteristics to observations. Here, the model is a description of the relationship between teams' abilities and their performances. Remember that we describe teams' abilities in terms of their talent numbers $t_1, \ldots, t_{30}$ and their performances using the season winning fractions $p_1, \ldots, p_{30}$. A model describes how the ts are linked to the ps.

Given the current parity of baseball teams and the movement of a large number of free-agents between teams, it might be reasonable to think that all baseball teams have roughly the same talent. If that is true, then all teams would have an average ability, and the talent number for each team would be 0. If this "mediocrity model" is true, then the observed differences in season winning percentages are solely due to good and bad luck. If you believe this model, then any team in 1997 had the same chance of winning the baseball championship. The Marlins just had the best luck, and that's why they won.

If this model is correct, then it would be easy to simulate a baseball season. Suppose any two teams play, say the Phillies and the Marlins. Since they are of equal abilities, then the probability the Phillies win the game is .5—this would be true for any other pair of teams. A complete season could be simulated by a sequence of coin flips, where each coin flip corresponds to the outcome of a single game.

Suppose that we do this simulation many times and keep track of the season winning fractions p for all teams. One team playing a season is analogous to flipping a fair coin 162 times and keeping track of the fraction of heads. Since we expect each team to win 50 percent of its games, a standard formula in statistics tells us that the season winning fractions for many teams will be normal shaped with mean .5 and standard deviation calculated as follows:

$$\sqrt{.5\,(.5)\,/\,162} = .0393$$

To see if this is a reasonable model, we compared the above distribution of winning fractions to the actual winning fractions that we observed for the recent ten-year period. Recall that the standard deviation of this distribution of actual team performances was estimated to be .0626. This standard deviation (.0626) is much larger than the standard deviation that would be predicted if the mediocrity model was true (.0393). Since this model doesn't explain the variation in winning percentages between teams, we reject it. Modern baseball teams do appear to have different abilities. This might seem to be a pretty obvious statement, but it illustrates how we state a model and how we can check if the model is a reasonable description of baseball competition. Even when a conclusion seems intuitively obvious, it is important to validate it with data before proceeding.

A Normal Model for Abilities

So baseball teams have different abilities. Remember, we describe a team's ability by a number t which we call the team's talent. There are 30 talent numbers that we write as the symbols $t_1, \ldots, t_{30}$. We saw in the previous section that it is inappropriate to assume that all of these talents are equal to 0. A model will tell us how these 30 team abilities can be different.

Recall that the performances of teams across different seasons are well described by a normal curve. Most teams have a winning fraction p that is in the neighborhood of .5, and a relatively small number of teams have poor or great winning fractions. It is reasonable to think that teams' abilities are also described by a normal bell-shaped curve. If a fan thinks about the quality of the current group of 30 baseball teams, then he or she will likely view most teams as "close to average" and think that there are only a few teams that are blessed with superior players and only a small number of teams that are rebuilding with young players. So we suppose that the team talents $t_1, \ldots, t_{30}$ have a normal pattern. We center this normal curve about the value 0, since we are assuming that an "average" team has a talent $t = 0$. Figure 11-9 displays this normal curve model for team abilities.

This normal model for the talents is centered about the mean value 0, which corresponds to an average team. The standard deviation of this curve tells us about the spread of team abilities. We know teams have different abilities, but can we figure out *how* different? In other words, how can we find the standard deviation of this normal curve? We choose a standard deviation so that the winning season percentages predicted from this normal ability model match the pattern of baseball winning season percentages that we saw in Figure 11-5. (We will shortly describe how the talent numbers determine who wins and loses individual baseball games.) How we actually find this standard deviation is a bit complicated. But it turns out that if we let the ability distribution have a standard deviation of .19, then the season performances (the p) that are predicted from this model match very well the observed season performances in Figure 11-5.

Recall that the standard deviation of the season proportions from recent years was .066. This variation in the teams' winning proportions is due to two factors. First, teams have different abilities, and the variation in these abilities is measured by the standard deviation of the normal curve of the team talents. But this variation in team abilities doesn't explain all of the variation in season winning proportions. The second factor is chance variability, which is analogous to the variation that we see in the number of heads when we toss 20 coins repeatedly. Teams perform well or poorly during a season due to different abilities, but also due to luck.

Weak, Average, and Strong Teams

Now that we have a good model for describing team abilities, we can use the model to group teams into meaningful categories. How does one define an "excellent" team? There are many ways to think of excellent teams, but we'll define them in a simple and somewhat arbitrary way. These are the teams that are in the top 10 percent with respect to team ability. Likewise a "bad" team is one that is in the bottom 10 percent of all the team talents. We'll define an "average" team that is in the middle 30 percent of the distribution. That leaves two final categories, "poor" and "good." Figure 11-9 shows where the different types of teams

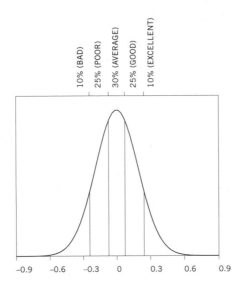

FIGURE 11-9   Normal curve model for team abilities.

fall in the distribution of team abilities. Table 11-2 gives the cutoffs for the different type of teams.

So an "excellent" team, one that is among the top 10 percent, has a talent number larger than +.24. A "good" team is one that has an ability between .07 and .24. "Average," "poor," and "bad" teams are defined in a similar way.

A Model for Playing a Season

We're discussing how to model baseball competition, and we have focused on how to model team abilities. But the model isn't complete. Given the team talents, we have to model the actual competition between teams in a 162-game regular season.

Bradley and Terry thought of a simple way of modeling a competition between a set of players or teams. For simplicity, suppose that there are four teams in the competition, which we'll call A, B, C, and D. We assign talents to the teams as given in Table 11-3. Under our normal model for talents, teams A and D have average ability, team B has below-average ability, and team C has the most talent.

We convert these talents to positive numbers, called strengths, by taking the exponential of each value. For example, we convert the talent number $t = 0$ to the strength number as follows:

$$s = e^0 = 1$$

Then we convert the talent value $t = -.1$ to

$$s = e^{-.1} = .9$$

Category	Percentiles of Ability Distribution	Team Talent
Bad	0–10	less than −.24
Poor	10–35	−.24 to −.07
Average	35–65	−.07 to .07
Good	65–90	.07 to .24
Excellent	90–100	larger than .24

TABLE 11-2   Five Categories of Ability of Baseball Teams

Team	A	B	C	D
Talent (t)	0	−0.1	0.4	0

TABLE 11-3   Talent Numbers Assigned to Four Teams

Team	A	B	C	D
Strength (s)	1	0.9	1.5	1

TABLE 11-4   Strength Numbers for Four Teams

(Note that e is a special mathematical number that is approximately equal to 2.78. So when we write $e^{-.1}$ we are taking the number 2.78 to the $-.1$ power.) If we do this exponential operation to all the talent numbers, we get the strengths as shown in Table 11-4.

We use the strength values to compute the probability that one team will defeat another team in a single game. Suppose two teams, say A and B, play one game. The chance that team A defeats team B is given by the following formula:

$$\text{Pr}(\textit{team A defeats team B}) = \frac{\textit{strength of team A}}{\textit{strength of team A + strength of team B}}$$

Here, team A has strength 1, B has strength .9, so the probability that A wins the game is as follows:

$$\text{Pr}(\textit{team A defeats team B}) = \frac{1}{1 + .9}$$

We can use the strength numbers to find the probability that any team defeats any other team. So the probability that team C (with strength 1.5) defeats D (with strength 1) is:

$$\frac{1.5}{1.5 + 1} = .6$$

What if two teams have equal strengths? Note that teams A and D both have strengths of 1. The chance that A defeats D is:

$$\frac{1}{1 + 1} = .5$$

which makes sense.

Simulating a Season

We now have a complete description of a model for modern baseball competition. Teams have different abilities, and we describe these abilities by means of a normal curve. Once we know the talent numbers for all of the teams, we can compute strength numbers for the teams, and these strength numbers are used, in the Bradley-Terry model, to compute the probability that one team will defeat another team in a single game.

Using this competition model, we can use random numbers to simulate a baseball season. We first choose abilities for the teams at random from the normal curve ability distribution. We then can play all of the games of the baseball season using probabilities given by the Bradley-Terry formula and a random spinner. To show how this simulation works, we'll step through a single simulation of the American League baseball season.

Simulating an American League Season

Let's focus on a hypothetical American League baseball season. The American League currently consists of 14 teams, arranged in the East, Central, and West divisions.

The first step in this simulation is to assign random abilities to these teams. Our model for the team talents is a normal curve with a mean 0 and standard deviation .19. We randomly select 14 numbers from this normal distribution and assign them to the teams. Table 11-5 lists the teams and their randomly assigned abilities.

The particular assignment of abilities to teams might look strange to the baseball fan, since they don't correspond to the current strengths of the teams. For example, Kansas City has a higher talent then Cleveland, which would seem very surprising to the 1999 fan. We could have assigned abilities based on our knowledge of the strengths and weaknesses of the individual teams. But what is important here is the spread of the talent numbers assigned. The spread of assigned team talents mimics the spread of abilities in modern-day baseball competition.

Also note that, since the talent numbers are given, we now know who should win each division title. Boston, Chicago and Anaheim have the largest abilities in their respective divisions, so they should win their divisions. Also, since Anaheim has the largest assigned talent, this team should be the American League representative in the World Series.

EAST DIVISION		CENTRAL DIVISION		WEST DIVISION	
Team	*Talent*	*Team*	*Talent*	*Team*	*Talent*
Tampa Bay	−0.288	Detroit	−0.128	Anaheim	0.140
Toronto	−0.223	Kansas City	−0.010	Texas	−0.010
Baltimore	0.030	Cleveland	−0.073	Oakland	0.095
Boston	0.086	Chicago	0.077	Seattle	−0.094
New York	0.039	Minnesota	−0.139		

TABLE 11-5   Randomly Assigned Talent Numbers for the American League Teams

Given these ability numbers, we compute strength numbers for all of the teams. For example, see from the Table 11-5 that Tampa Bay and Toronto had respective talent numbers of $-.288$ and $-.223$. So their respective strength numbers are:

$$s = e^{-.288} = .75, \ s = e^{-.233} = .80$$

These numbers are listed in Table 11-6.

Now we can proceed with the simulation of the baseball season. We'll use a baseball schedule close to the actual schedule used in Major League Baseball. In this schedule, each team will play each of the other teams in its division 14 times and play each team in the other two divisions 12 times. A slight adjustment is made to the schedule so that every team plays a total of 162 games.

It's now opening day in our simulated season, and New York is playing Toronto. The strength numbers for these two teams are respectively 1.04 and .80, and so the probability that New York wins this game is $1.04/(1.04 + .80) = .565$. The probability that Toronto wins this game is $1 - .565 = .435$. We can play this game using a random spinner. In Figure 11-10, we have drawn a circle where the areas of the two regions correspond to the probabilities that New York

EAST DIVISION		CENTRAL DIVISION		WEST DIVISION	
Team	*Strength*	*Team*	*Strength*	*Team*	*Strength*
Tampa Bay	0.75	Detroit	0.88	Anaheim	1.15
Toronto	0.80	Kansas City	0.99	Texas	0.99
Baltimore	1.03	Cleveland	0.93	Oakland	1.10
Boston	1.09	Chicago	1.08	Seattle	0.91
New York	1.04	Minnesota	0.87		

TABLE 11-6   Randomly Assigned Strength Numbers for the American League Teams

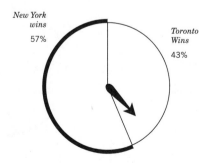

FIGURE 11-10   Spinner for simulating the result of a single baseball game.

and Toronto win. Imagine spinning an arrow which is equally likely to land anywhere around the circle. If the arrow lands in the New York region, New York wins the game; otherwise, Toronto wins. Note from the figure that the spinner lands in the Toronto region, so Toronto wins this particular game.

Other games are simulated in this same manner. Figure 11-11 shows the simulation for four other opening-day games. For each game, we construct a spinner divided into two regions, where the areas of the regions correspond to the probabilities that each team wins the game. (We compute the probabilities from the two teams' strength numbers.) Then we spin the spinner, and the location of the arrow tells us who won the game. We see that the winners of these games on this day were Boston, Minnesota, Chicago and Anaheim.

We continue this process until we have played a complete 162-game season for these fourteen American League teams. All of the games are played using win probabilities based on the team abilities that were assigned at the beginning of the season. How did our teams do this season? The final standings of the teams are shown in Table 11-7.

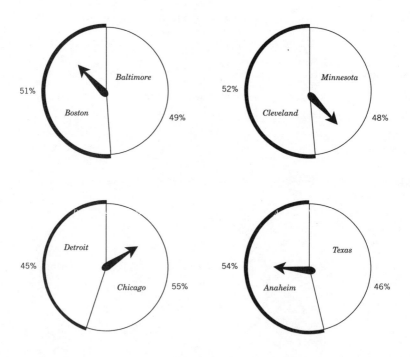

FIGURE 11-11 Random spinners for playing four games on opening day.

EAST DIVISION

Team	W	L	p
Boston	88	74	0.543
New York	84	78	0.518
Baltimore	77	85	0.475
Toronto	66	96	0.407
Tampa Bay	63	99	0.389

CENTRAL DIVISION

Team	W	L	p
Detroit	90	72	0.556
Chicago	84	78	0.518
Cleveland	82	80	0.506
Kansas City	78	84	0.482
Minnesota	74	88	0.457

WEST DIVISION

Team	W	L	p
Anaheim	101	61	0.624
Texas	94	68	0.580
Oakland	80	82	0.494
Seattle	73	89	0.451

TABLE 11-7   Results of One Simulated Baseball Season

The results of the simulated season may surprise you. Let's focus on the American League West. Looking at the abilities of the four teams, Anaheim and Oakland had above-average abilities (with positive talent values), Texas had average ability (talent close to 0), and Seattle had an ability in the below-average range (negative talent). Although Anaheim and Oakland had similar abilities, Anaheim won the division title very easily—they finished with a 7-game lead over second-place Texas. Anaheim and Texas played much better than their abilities, while Oakland and Seattle played worse then their abilities. So we see significant differences between the teams' abilities and performances for this particular season.

To see how the abilities for all 14 teams are related to their season performances, Figure 11-12 displays a scatterplot of the values of the talents t and the season win fractions p. We see a positive drift in the plot, which indicates that there is a moderate positive relationship between teams' abilities and performances. We have placed a best fitting line on the scatterplot. Points above the line correspond to teams that played better in the season than expected, and points below the line correspond to "disappointing" team performances. In particular, we have labeled points corresponding to Anaheim, Texas, and Detroit, who had

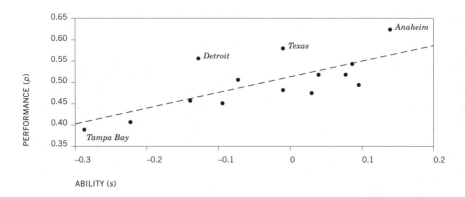

FIGURE 11-12 Abilities and simulated season performances for 14 American League teams.

better-than-expected years. We have also labeled one point under the line that corresponds to Tampa Bay, which had a disappointing season. This graph illustrates that the relative standing of the teams' abilities will generally be different from the relative standing of the teams' performances during a 162-game season. To illustrate this, note from the graph that Texas had the second-best record during the season, but there were a number of teams that had greater ability than Texas. But when the entire league as a whole is examined via the trend line, a reasonable relationship does exist between ability and performance.

Simulating Many American League Seasons

Above we simulated one baseball season between teams of the American League and found that the results were a bit surprising. Some teams with similar abilities had very different season performances, and the winners of the divisions were not necessarily the ones with the greatest abilities. But this one simulated season may have been a fluke. Perhaps we're members of the army of Yankee-haters and just happened to pull out one particular simulation where the Yankees had a particularly bad season.

To get a better understanding of the *pattern* of the relationship between team abilities and team performances, we can repeat this baseball simulation a large number of times. Remember, this simulation is a two-step process: first we generate a set of team talents (ts) from the normal curve ability model, then we play out a season of games using the Bradley-Terry model and a set of random spinners.

We repeat the American League season simulation a total of 1000 times. For each team and for each season, we keep track of two quantities: the team's ability given by its talent number t, and the team's winning percentage p for the

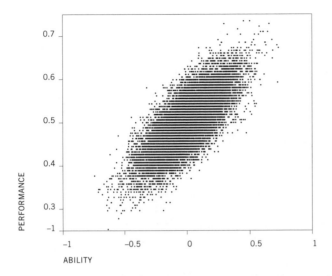

FIGURE 11-13 Scatterplot of abilities and simulated season performances for American League teams in 1000 seasons.

162-game season. Figure 11-13 displays a scatterplot of the team abilities against the team performances for all 14 teams playing 1000 seasons. There are several interesting features in this plot. First, there is a moderate positive relationship between the teams' abilities and the teams' performances. Teams with higher abilities tend to win a higher proportion of games during a season. But there is also a lot of scatter in this plot. This means that season performances of teams can be very different from their abilities.

To illustrate this last point, look at all of the teams with average abilities—that is, talent numbers close to 0. (This is the vertical line in Figure 11-13 passing through the value 0 on the team ability scale.) Figure 11-14 shows a boxplot of the season winning fractions of the truly average teams whose talent number is close to 0. We see from this boxplot that these teams had season winning fractions falling between .4 and .6. So it's possible for a team to have average talent ($t = 0$) and have very bad ($p = .4$) or very successful ($p = .6$) seasons.

Performances and Abilities of Different Types of Teams

Recall our categorization of teams of different abilities. The top 10 percent of the teams on the talent scale are considered "excellent," the teams between the percentiles 10 and 35 in talent are considered "good," the teams between the percentiles 35 and 65 are considered "average," and so on. Another way of thinking

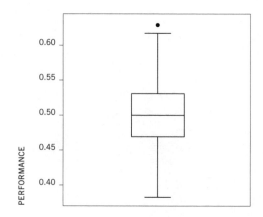

FIGURE 11-14 Boxplot of simulated season performances for American League teams of average ability where the talent number is close to 0.

about a team's ability is how it performs in the long run. Suppose that the team plays a very long sequence of games against random opponents that is much longer than the 162 games that a team plays during a major-league season. Then we define a long-run winning proportion P as the fraction of games that a team wins. For example, consider a team with a $P = .6$. This team will, in the very long run, win about 60 percent of its games. We'll see that this team can win fewer than 60 percent or more than 60 percent of the games during a 162-game season. But if this team were able to play thousands or even millions of games, the fraction or proportion of games won would be close to 60 percent. This team has above-average ability since its long-run winning fraction is over 50 percent.

If one knows the talent number t of a team, one can compute its long-run winning proportion P. Table 11-8 shows the long-run winning fractions of teams of different ability levels.

Category	Percentiles of Ability Distribution	Team Talent	Long-run Winning Proportion (p)
Bad	0 – 10	less than –.24	.000 – .442
Poor	10 – 35	–.24 to –.07	.442 – .483
Average	35 – 65	–.07 to .07	.483 – .517
Good	65 – 90	.07 to .24	.517 – .558
Excellent	90 – 100	larger than .24	.558 – 1.000

TABLE 11-8   Five Categories of Ability of Teams with Associated Team Talents and Long-run Winning Fractions

Suppose that we use the categorization of Table 11-8 to describe season performances. So an "excellent" season performance corresponds to a team that wins at least 55.8 percent of their games, which corresponds to a season record of 91–71 or better. A "good" performance means that a team wins between 51.7 percent and 55.8 percent of its games—in a 162-game season, this type of team will win have a record ranging from 84–78 to 90–72. Table 11-9 lists the five types of teams and what kind of seasons they will have.

Let's return to our simulation of 1000 American League seasons. For each team and each season, we can classify its ability depending on its talent number t. In addition, after the team has played its season, we can classify its season based on the proportion of wins p. Since there are 14 teams in the American League, each playing 1000 seasons, there are a total of $14 \times 1000 = 14,000$ team seasons. Table 11-10 classifies the performances and abilities for these teams by means of a two-way table. The rows of the table correspond to different team abilities, and the columns correspond to different team performances. Each table entry represents the number of teams having a specific ability and performance

Category	Winning Proportion (p)	Number of Games Won
Bad	.000 – .442	71 or fewer
Poor	.442 – .483	72 to 78
Average	.483 – .517	79 to 83
Good	.517 – .558	84 to 90
Excellent	.558 – 1.000	91 or more

TABLE 11-9   Five Categories of Performance of Teams in a 162-Game Season

PERFORMANCE

Ability	Bad	Poor	Average	Good	Excellent	Total
Bad	1015	299	56	6	0	1376
Poor	1090	1312	706	334	43	3485
Average	371	1146	1112	1063	350	4042
Good	47	374	716	1431	1054	3622
Excellent	0	16	85	336	1038	1475
Total	2523	3147	2675	3170	2485	14000

TABLE 11-10   Abilities (Rows) and Performance (Columns) of American League Teams in 1000 Simulated Seasons

level. To help understand this table, note that the count in the third row (Average Ability) and first column (Bad Performance) is 371. So there were 371 teams of average ability that had bad seasons. Looking across the same row, there were 1146 teams of average ability that had poor seasons, 1112 teams of average ability that had an average season, and so forth.

This two-way table is useful for seeing how teams of different abilities perform during a 162-game season. Suppose that we convert the table to row percentages by dividing each count by the total count of the corresponding row. The resulting table is shown in Table 11-11. Look at the first row—these numbers represent percentages of the teams with bad abilities. Of these bad teams, 73.8 percent had bad seasons, 21.7 percent had poor seasons, and 4.1 percent had average seasons. Also note that there are zeros in the Good and Excellent columns; these indicate that these bad teams never had good or excellent seasons. So teams with very weak abilities tend to play badly during a season. What about the poor teams? The possible performances of these teams is pretty spread out—31 percent of their performances fall into the category "bad," 38 percent "poor," 20 percent "average," and 10 percent "good." So it is possible (but not probable) that a poor team will have a good season. The performances of the teams with average abilities are the most spread out. These teams are equally likely to have poor, average, or good seasons. Also, these average teams have a plausible (9 percent) chance of having bad or excellent seasons.

The data in Table 11-11 can be used in a different manner. When we observe the results of a single baseball season, we're interested in what is learned about the team's ability. For example, suppose our team has a "good" season, that is, they win between 84 and 90 games. What can we say about the team's ability? We can answer these type of questions by converting the table of counts to col-

PERFORMANCE

Ability	Bad	Poor	Average	Good	Excellent
Bad	73.8	21.7	4.1	0	0
Poor	31.3	37.6	20.3	9.6	1.2
Average	9.2	28.4	27.5	26.3	8.7
Good	1.3	10.3	19.8	39.5	29.1
Excellent	0	1.1	5.8	22.8	70.4

TABLE 11-11   Performances of Simulated Teams of Different Ability Levels. Each Number Represents a Percentage of the Row

PERFORMANCE

Ability	Bad	Poor	Average	Good	Excellent
Bad	40.2	9.5	2.1	0	0
Poor	43.2	41.7	26.4	10.5	1.7
Average	14.7	36.4	41.6	33.5	14.1
Good	1.9	11.9	26.8	45.1	42.4
Excellent	0	0.5	3.2	10.6	41.8

TABLE 11-12   Performances of Simulated Teams of Different Performance Levels—Each Number Represents a Percentage of the Column

umn percentages—that is, we divide each count by the total in the correspon-ding column. Table 11-12 gives us insight into the abilities of the teams that have different types of seasons. To illustrate, look at the first column of the table, which corresponds to teams that had bad seasons. Of these teams, 40 percent were actually teams whose ability was categorized as "bad," 43 percent were "poor," 15 percent were "average," and 2 percent were "good." So it is likely that this team was a bad or poor team. What if our team has an average season? Does it mean that this team was average in ability? Looking at the third column of the table, corresponding to average, we see that there is a 42-percent chance that this team was actually average, and a 26-percent chance that the team was poor, the same 26-percent chance that the team was good, and relatively small chances that the team was bad or excellent in ability.

Simulating an Entire Season

Up to this point we've focused on what happens in an American League 162-game season with 14 teams, relating the teams' abilities with their season per-formances by use of a simulation experiment. But, as all baseball fans know, baseball really gets exciting when the regular season ends and the playoffs begin. At this point, a select group of teams get to continue in a series of playoffs, with the ultimate goal of winning the World Series.

We can extend the simulation we did earlier in this chapter to include all playoffs. As in the earlier simulation, we begin by simulating a set of abilities for all teams in the major leagues, including both the American and National Leagues. Then each team plays a complete 162-game season. At the end of the regular season, the division winners and wild card teams are found. Then the simulation can be used to play all of the post-regular-season series, concluding with the "best-of-7" World Series.

In investigating the "extended season," we simulated a total of 1000 complete baseball seasons. For each team and each season, its randomly generated ability and its season performance (wins and losses) were recorded. Also we recorded if the team achieved any of the following distinctions:

- The team won its respective division.

- The team was a wild card team for its league.

- The team won its pennant (was the winner of its league) and appeared in the World Series.

- The team won the World Series and was champion of baseball.

Table 11-13 summarizes what happened in these 1000 simulated baseball seasons. This table gives the number of teams of each ability level that reached various plateaus. Looking at the first row of the table, we see that there were a total of 2976 bad teams in all of the simulations. Of these 2976 teams with bad ability, only 15 teams won a division, and one team was a wild card team. None of these teams ever won a pennant or a World Series. So it is virtually impossible for a bad team to win a World Series. The performance of the teams of average ability is more interesting. Of the 8600 teams of average ability in the simulation, 1140 (13 percent) won their divisions, 483 (6 percent) were wild card teams, 839 (3 percent) won their pennants, and 116 (1 percent) won the World Series. So it's possible, but not likely, that these average teams will achieve success in a season. What about the success of the excellent teams that represent the top 10 percent of all teams? Of the 3031 teams of this type, 1975 (65 percent) won their division, 453 (15 percent) were wild card teams, 840 (28 percent) won

PLAYOFF PERFORMANCE

Ability	Won Division	Wild Card Team	Won Pennant	Won World Series	Won Nothing	Total
Bad	15	1	0	0	2960	2976
Poor	278	110	53	17	7194	7582
Average	1140	483	268	116	6977	8600
Good	2592	953	839	397	4266	7811
Excellent	1975	453	840	470	603	3031

TABLE 11-13   Playoff Performances of Simulated Teams of Different Ability Levels

their pennants, and 470 (16 percent) won the World Series. These top teams will achieve success, but perhaps not at the high rate that one would expect. Table 11-14 shows these teams' chances of reaching these plateaus for teams of all ability levels. Figure 11-15 displays these probabilities using line graphs.

Again let's turn this logic around. Suppose a team wins the World Series—is it reasonable to call this team the "best team in baseball"? In our simulation, there were 1000 World Series winners. Looking at the fourth column of Table 11-13, we see that 470 (47 percent) of these teams had excellent ability, 397 (40 percent) were good, 116 (12 percent) were average, and 17 (2 percent) were poor. So the answer to our question is no. The chance that a great team, a team of excellent ability, won the World Series is under 50 percent. It *is* likely that one

PROBABILITY THE TEAM...

Ability	Won Division	Wild Card Team	Won Pennant	Won World Series	Won Nothing
Bad	0.01	0	0	0	0.99
Poor	0.04	0.01	0.01	0	0.95
Average	0.13	0.06	0.03	0.01	0.81
Good	0.33	0.12	0.11	0.05	0.55
Excellent	0.65	0.15	0.28	0.16	0.2
The Best	0.76	0.13	0.35	0.21	0.21

TABLE 11-14 The Probability a Team of Different Ability Levels Reaches Different Playoff Levels

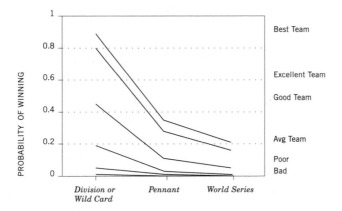

FIGURE 11-15 Graph of probabilities of reaching different plateaus for teams of different ability levels.

or the other of the two (a good or excellent team) wins this contest, and it is possible (but not likely) for an average and even a poor team to accomplish this feat. By looking at the Won Division and Won Pennant columns of Table 11-13, we can check into the abilities of teams that won the division and won the pennant, respectively.

Let's answer one final question. The title we gave this chapter was, "Did the Best Team Win?" In other words, is it likely that the best team—that is, the team with the greatest ability among all 30—will win the World Series? Using the simulation, this question is easy to answer. For each simulated season, after we simulate the talents (the ts) for all teams, we find the particular team with the largest talent. We then record if this team won the World Series that year. The results of the success of the best team in these 1000 seasons is presented in the last row of Table 11-14. For the 1000 seasons, the best team—the team with the highest value of t—won the World Series 213 times. So the chance that the best team wins the World Series is about 21 percent. We also find that this best team wins a division title with a probability of 76 percent, is a wild card team with a probability of 13 percent, and wins the pennant with a probability of 35 percent. So this best team will very likely get into the baseball playoffs, but it has a modest chance of actually winning the World Series. To put it another way: the cream won't generally rise to the top.

Chance

The point of this chapter is to relate the abilities of major-league teams with their performances during a baseball season. We measure a team's ability by a talent number t. Teams possess different talents, that is, different abilities, and we model the distribution of abilities of all teams by a bell-shaped curve. By choosing a reasonable value for the spread (the standard deviation) of this curve, the observed season winning fractions (p), predicted using the model for different teams, will match the actual winning percentages that we observe for major-league teams over the last 20 years. Our model for baseball competition consists of this normal curve model for the teams' abilities and a simple coin-tossing model (the Bradley-Terry model) for describing the results of games between teams of different abilities. It is important to stress that the model seems reasonable in that it appears to predict well the observed season results for the 30 major-league teams.

By using the model, we simulated a large number of baseball seasons. In each simulation, we select at random a set of team talents and then use these talents to play a 162-game baseball season and the playoffs. We use the simulation results to connect the team abilities (that we don't know in real life) and their season performances. What we learned is that teams with high abilities tend to perform better than teams of low abilities. But it is pretty common for good teams to have average or worse seasons—likewise, mediocre teams can have good seasons. Probably the most surprising result, from a typical fan's perspective, is the range of abilities found in teams that win the World Series. Over half of the World Series winners are in the "good," "average," or "bad" ability categories. So we shouldn't be too surprised when a team like the Florida Marlins in 1997 wins the World Series over a clearly superior team like the Atlanta Braves. Also, this study should be encouraging to a Philadelphia Phillies fan— even if the team doesn't possess great talent (as is usually the case!), the Phillies have a reasonable chance for a good season, even to the point of winning the fall classic.

In other words, we've shown that chance variability has a lot to do with teams' performances during a season. When a team reaches a certain pinnacle such as the World Series, sportswriters and fans will offer a thousand explanations why this team performed so well. Any good performance has to have a cause—perhaps a few ballplayers got "hot" or performed at a higher level than expected, or perhaps a few players on the opposing team experienced slumps. Maybe the umpires made a number of questionable calls which influenced the outcome. Many things can happen during a game that cannot be explained easily and yet influence the final result. These "things" include good or bad pitches, the locations of balls hit in the infield and outfield, good and bad defensive or base-running plays, the weather conditions, and so on. We can lump all of these events into a broad category called "chance occurrences." But lumping them together and giving them this name does not diminish their importance: Whether we like it or not, chance events have a big effect on the patterns of wins and losses that we observe.

POSTGAME COMMENTS (A Brief Afterword)

CHAPTER 12

Baseball is a fascinating game for the statistical analyst. On the surface, it appears so simple and limited. But the more closely one studies the game, the more, it seems, there is to know. For us it is like driving toward the horizon, then at the top of that last hill finding a valley on the other side, with more hills and another horizon.

The chapters in this book have examined some aspects of baseball statistics that we've found particularly interesting over the years. We make no claim that this is the last word on these issues, nor do we make any claims for completeness. As we write, there are thousands of fans taking fresh new looks at baseball and the usefulness of its statistical underpinnings. Undoubtedly, refinements and advances will be made. Our book is not, for example, a complete guide to sabermetrics. Many issues of interest to sabermetricians and their disciples are not covered, or are examined superficially.

This book is best approached as a loosely connected collection of quantitative essays on baseball statistics. Each essay examines a topic in baseball in the way a professional statistician might approach the issue in any other field (health, business, technology) using the data at hand. In some cases, the work presented was original research performed by us, the authors, while in other cases the essay described work performed by others whose efforts we have found particularly enlightening. Hopefully, we have brought to your notice some older research that has not received the attention it deserves, and provided a new perspective on work with which you're already familiar.

But there is a common thread that runs through the chapters, and it can be summarized as the role of *chance* in baseball. A Major League Baseball season (and indeed a single batter's swing or the whole history of the sport) is a process that is at the mercy of chance. Chance affects all baseball events, from the outcomes of individual at-bats to the awarding of World Championships. While chance occurrences may seem to defy description and analysis, to make explanation and prediction impossible, we believe chance can be mastered—or at least tamed. This is the role of statistics as it is applied in business and industry. Statistics involves not just enumerating and summarizing data, but attempting to extract the underlying truth that is obscured by the fog of chance.

If there is one lesson the reader should take from this book, it is that baseball data are like observations in experiments. They are the best measures we have, but they are not exact with respect to the underlying process. The result of each at-bat is the culmination of many factors, but even if all are held constant, the difference in the batter's reflexes or reactions for one tiny fraction of a second, or one tiny fraction of an inch one way or the other, may make the difference between a strikeout and a home run. We cannot (and would not wish to) alter these elements of chance in the game. But when we analyze baseball data, we should attempt to take them into account.

Several chapters in this book have presented ways in which we can gauge the influence of chance on baseball. We always assume that the winner of the World Series is the best team in baseball, but in Chapter 11 we found that the role of chance in the game gives an inferior team a significant shot at winning the series. By inferior, we mean that the championship team's abilities are not as good as those of some other teams. It may have had great performances that year, but those performances were better than expected.

The trick is to find the signal in the noise. At times, this can be difficult. Chapter 4 examined ways of detecting situational effects in batting data. Baseball announcers present figures on batting averages in the day, at night, on turf, on grass, against left-handers, against right-handers, at home or away, and so forth. These figures are presented as facts. And they are factual observations of what has occurred. But generally they are not valid statements about how these situations affect ability. Chapter 4 demonstrated how certain effects were large enough to be beyond the realm of chance, while others were indistinguishable from chance.

Similarly, baseball announcers are quick to highlight notable batting streaks and discuss certain hitters as being hot or cold. But how much of this is due to chance as opposed to the possibility that a batter really has an ability which fluctuates depending on recent success? Chapter 5 examined one player (Todd Zeile) noted for his streakiness. While evidence was found to support the possibility that Zeile truly had a streaky kind of ability, it was still difficult to rule out chance as the primary cause. Generally, while some players may exhibit some streaky batting behavior, one should be pretty doubtful of labeling someone a streaky hitter. Chance is a very powerful force in creating streaks.

In Chapters 6 and 7, we examined different ways of measuring player contributions to run production. The primary metric used to determine how well these measures worked was Root Mean Squared Error (RMSE) as it related to team run production. RMSE is a measure of how much chance remains in a prediction. The RMSEs for standard measures such as Batting Average were relatively large, leaving chance a major element in the prediction of runs produced. Newer measures such as Runs Created and Linear Weights had much lower RMSEs; their predictions greatly reduced the element of chance in predictions of run production. The capability of a measure to control or limit chance is a major factor in making the measure a useful statistical tool.

Simulation models attempt to incorporate the element of chance. In Chapter 8, we described a very basic simulation model that used the rules of probability to model the chance elements in games. The model was found to match closely the variability of runs scored in an inning. The structure and behavior of the model provided some theoretical support for the Runs Created model, which was developed totally (almost) from intuition. An interesting feature of this simulation was its construction from formulas of probability that made it possible to produce predictions without the need for great numbers of computer replays.

Chapter 9 found a use for chance as a measure in itself. Where most measures for player value focus on run production or run prevention, two related measures of player contributions were described based on how much a player increases or decreases his team's chance of winning. These measures provide the next step in the evolution of baseball metrics, going beyond counts of individual events to measures of run production and on to measuring the ultimate goal, winning.

While chance is a major element of baseball (and everything else), it is not powerful enough to make baseball data arbitrary. We just have to be a bit more wise in our analysis of the data to understand the degree of chance's control and how we may allow for its influence in our understanding. A direct blunt analytical approach (using simple averages) will often be satisfactory, but it can also prove to be deceptive.

Or, to put the same bit of advice in baseball terms, look for the fast ball, but watch out for the curve.

Bibliography

Albert, Jim (1994), "Exploring Baseball Hitting Data: What About Those Breakdown Statistics?" In: *Journal of the American Statistical Association*, 89, 427, pp. 1066–1074.

Bennett, Jay (1993), "Did Shoeless Joe Jackson Throw the 1919 World Series?" In: *The American Statistician*, 47, 4, pp. 241–250.

Bennett, Jay (1994), "MVP, LVP, and PGP: A Statistical Analysis of Toronto in the World Series." In: *1994 Proceedings of the Section on Statistics in Sports*, American Statistical Association, pp. 66–71.

Bennett, Jay (ed.) (1998), *Statistics in Sport*. London: Arnold.

Bennett, Jay M. and Flueck, John A. (1983), "An Evaluation of Major League Baseball Offensive Performance Models." In: *The American Statistician*, 37 (1), pp. 76–82.

Bennett, Jay M. and Flueck, John A. (1984), "Player Game Percentage." In: *1984 Proceedings of the Social Statistics Section*, American Statistical Association, pp. 378–380.

Berry, Donald A. (1996), *Statistics: A Bayesian Perspective*. Belmont, CA: Wadsworth Publishing.

Berry, Scott (1999), "How Many Will Big Mac and Sammy Hit in '99?" In: *Chance*, 12, pp. 51–55.

Boswell, Thomas (1981), "Win Arguments and Impress Your Friends with This Startling New Baseball Stat: Welcome to the World of Total Average." In: *Inside Sports*, January 31, 1981.

Bradley, R. A. and Terry, M. E. (1952), "Rank Analysis of Incomplete Block Designs, I, The Method of Paired Comparisons." In: *Biometrika*, 39, pp. 324–345.

Carroll, Bob, Palmer, Pete, and Thorn, John (1988), *The Hidden Game of Football*. New York: Warner Books.

Chieger, Bob (ed.) (1983), *Voices of Baseball*. New York: New American Library.

Codell, Barry F. (1979), "The Base-Out Percentage: Baseball's Newest Yardstick." In: *Baseball Research Journal*, pp. 35–39.

Cook, Earnshaw (1977), "An Analysis Of Baseball As a Game of Chance by the Monte Carlo Method." In: *Optimal Strategies in Sports* (edited by S.P. Ladany and R.E. Machol), pp. 50–54. New York: North-Holland.

Cook, Earnshaw (in collaboration with Garner, Wendell R.) (1966), *Percentage Baseball*. Second Edition, Cambridge, MA: The MIT Press.

Cover, Thomas M. and Keilers, Carroll W. (1977), "An Offensive Earned Run Average for Baseball." In: *Operations Research*, 25, pp. 729–740.

Cramer, Richard D. and Palmer, Pete (1974), "The Batter's Run Average (B.R.A.)." In: *Baseball Research Journal*, pp. 50–57.

D'Esopo, D. A. and Lefkowitz, B. (1977), "The Distribution of Runs in the Game of Baseball." In: *Optimal Strategies in Sports* (edited by S. P. Ladany and R. E. Machol), pp. 55–62. New York: North-Holland.

Dickson, Paul (1989), *The Dickson Baseball Dictionary*. New York: Avon Books.

Dickson, Paul (1991), *Baseball's Greatest Quotations*. New York: Harper Collins.

Friedman, Arthur (1978), *The World of Sports Statistics*. New York: Atheneum.

Hastings, Kevin (1999), "Building a Baseball Simulation Game." In: *Chance*, 12, 1, pp. 32–37.

James, Bill (1984), *The Bill James Baseball Abstract 1984*. New York: Ballantine Books.

James, Bill (1994), *The Politics of Glory*. New York: Macmillan.

James, Bill (1997), *The Bill James Guide to Baseball Managers*. New York: Scribners.

James, Bill, Albert, Jim, and Stern, Hal S. (1993), "Answering Questions about Baseball Using Statistics." In: *Chance*, 6, 2, pp. 17–22, 30.

Ladany, S. P. and Machol, R. E. (eds.) (1977), *Optimal Strategies in Sports*. New York: North-Holland.

Lindsey, G. R. (1959), "Statistical Data Useful for the Operation of a Baseball Team." In: *Operations Research*, 7, 2, pp. 197–207.

Lindsey, G. R. (1961), "The Progress of the Score During a Baseball Game." In: *Journal of the American Statistical Association*, 56, 295, pp. 703–728.

Lindsey, George R. (1963) "An Investigation of Strategies in Baseball." In: *Operations Research*, 11, 4, pp. 477–501.

Malcolm, Don, Hanke, Brock J., Adams, Ken, and Walker, G. Jay, (1999), *The 1999 Big Bad Baseball Annual*. Chicago: Masters Press.

Mills, Eldon G. and Mills, Harlan D. (1970), *Player Win Averages*. South Brunswick, NJ: A. S. Barnes.

Moore, David S. (1997), *Statistics: Concepts and Controversies*. Fifth Edition, New York: W. H. Freeman and Company.

Neft, David S., and Cohen, Richard M. (1997), *The Sports Encyclopedia: Baseball*. New York: St. Martin's Press.

Schell, Michael (1999), *Baseball's All-Time Best Hitters: How Statistics Can Level the Playing Field*. Princeton, NJ: Princeton University Press.

Siwoff, Seymour (ed.) (1999), *The Book of Baseball Records*. New York: Elias Sports Bureau.

Siwoff, Seymour, Hirdt, Steve, and Hirdt, Peter (1985), *The 1985 Elias Baseball Analyst*. New York: Collier Books.

Siwoff, Seymour, Hirdt, Steve, and Hirdt, Peter (1987), *The 1987 Elias Baseball Analyst*. New York: Collier Books.

Siwoff, Seymour, Hirdt, Steve, and Hirdt, Peter (1988), *The 1988 Elias Baseball Analyst*. New York: Collier Books.

Siwoff, Seymour, Hirdt, Steve, Hirdt, Tom, and Hirdt, Peter (1989), *The 1989 Elias Baseball Analyst*. New York: Collier Books.

Siwoff, Seymour, Hirdt, Steve, Hirdt, Tom, and Hirdt, Peter (1991), *The 1991 Elias Baseball Analyst*. New York: Fireside Books.

Siwoff, Seymour, Hirdt, Steve, Hirdt, Tom, and Hirdt, Peter (1993), *The 1993 Elias Baseball Analyst*. New York: Fireside Books.

The Sporting News, "Baseball's 25 Greatest Moments." October 18, 1999.

STATS Player Profiles 1999. Stats Inc.

Stern, Hal S. (1997), "How Accurately Can Sports Outcomes Be Predicted?" In: *Chance*, 10, 4, pp. 19–23.

Thorn, John and Palmer, Pete (1985), *The Hidden Game of Baseball*. New York: Doubleday.

Thorn, John and Palmer, Pete (eds.), (1989), *Total Baseball*. New York: Warner Books.

Verducci, Tom (1999), "A Game for Unlikely Heroes." In: *Sports Illustrated*, November, 29, 1999.

Warrack, Giles (1995), "The Great Streak." In: *Chance*, 8, 3, pp. 41–43, 60.

Zoss, Joel and Bowman, John (1989), *Diamonds in the Rough: The Untold History of Baseball*. New York: Macmillan.